HUMAN AND ORGANIZATIONAL FACTORS IN NUCLEAR SAFETY

Human and Organizational Factors in Nuclear Safety

The French Approach to Safety Assessments

Grégory Rolina
École des Mines de Paris, France

CRC Press
Taylor & Francis Group
Boca Raton London New York Leiden

CRC Press is an imprint of the
Taylor & Francis Group, an informa business

A BALKEMA BOOK

Cover photo information:
Central Nuclear Almirante Álvaro Alberto (Angra dos Reis, RJ, Brazil), Operated by
Eletrobras Eletronuclear. Copyright Eletrobras Eletronuclear.

CRC Press
Taylor & Francis Group
6000 Broken Sound Parkway NW, Suite 300
Boca Raton, FL 33487-2742

First issued in paperback 2017

ISBN: 978-1-138-00035-3 (hbk)
ISBN: 978-1-138-07164-3 (pbk)

Originally published in French as: Sûreté nucléaire et facteurs humains,
La fabrique française de l'expertise
© 2009 Transvalor – Presses des Mines, Paris, France

English edition: 'Human and Organizational Factors in Nuclear Safety',
CRC Press / Balkema, Taylor & Francis Group, an informa business
© 2013 Taylor & Francis Group, London, UK

Typeset by MPS Limited, Chennai, India
Translation: Provence Traduction, Stephen Schwanbeck

Library of Congress Cataloging-in-Publication Data

Rolina, Grégory.
   [Sûreté nucléaire et facteurs humains. English]
   Human and organizational factors in nuclear safety : the French approach to safety
assessments / Grégory Rolina, École des mines de Paris, France.
      pages cm
   Includes bibliographical references and index.
   ISBN 978-1-138-00035-3 (hardback : alk. paper) — ISBN 978-0-203-76887-7 (eBook)
   1. Nuclear facilities—Security measures—France. 2. Nuclear power plants—Human factors—
France. I. Title.
   TK9152.R6513 2013
   621.48'30684—dc23
                                                                        2013030725

Visit the Taylor & Francis Web site at
http://www.taylorandfrancis.com

and the CRC Press Web site at
http://www.crcpress.com

Contents

Preface to the English Version

Four years have elapsed since the publication, in France, of Grégory Rolina's book detailing the findings of his research on how safety assessments are prepared at IRSN, the French public institution in charge of assessing nuclear and radiological risks on behalf of the French public authorities.

The fact that global news coverage has been dominated for the past two years by the Fukushima Daiichi accident testifies to the value of this book, which is illustrated with examples that allow the reader to understand, in a very concrete way, the French method of constructing nuclear safety assessments.

Indeed, without downplaying the extraordinary natural causes behind this industrial and environmental disaster, learning all the lessons from this accident requires a detailed analysis of the mechanisms used to govern the risks and respond to the emergency.

By reconstructing several safety assessments conducted by IRSN, Grégory Rolina presents just such a type of analysis. He brings to light the cultural particularities of the French model of nuclear risk governance, which is underpinned by safety reference frameworks that take into account the particularities of each individual licensee and facility and requires close, ongoing technical dialogue between licensees and IRSN. A critical comparison of the models used in various countries appears today to be vital to improving the decision-making mechanisms for ensuring the safety of nuclear facilities.

The field of expertise that Grégory Rolina examines in his book is that of the human and organizational factors of nuclear and radiological risks. The Fukushima accident highlights the importance of expanding knowledge in this field. It is for this very reason that, in 2012, IRSN created a human and social sciences research laboratory. By helping to imagine the unimaginable, the research conducted in this field of expertise will, like this book, contribute to further improve global nuclear safety.

I am therefore delighted that it is now available in English, and thus accessible to a wider audience.

Fontenay-aux-Roses – May 17, 2013

Jacques Repussard,

IRSN Director General

Preface

It is a great pleasure for me to preface this book presenting the results of research on nuclear facility safety assessments conducted in France, a very special area of expertise. Through the critical eye of completely independent experts, assessments examine measures implemented by facility licensees to control radiological and nuclear risks, and serve to support the decision-making process of the French nuclear safety authority, the ASN. They are conducted by the French Institute for Radiological Protection and Nuclear Safety (IRSN), a public institution. Safety assessments, site inspections and analysis of operating experience feedback form the three key actions taken by public authorities to establish a comprehensive view of safety at nuclear facilities in France.

Written by a researcher from outside IRSN, this book provides a rigorous and detailed description of IRSN's practices in the area of nuclear safety. By describing the decisive moments and key operations in the assessment process, the book helps specialists and non-specialists alike to understand the steps involved in safety assessment, and the ways in which it contributes to safety at French nuclear facilities. Publication of this research, carried out in collaboration with a team of experts from IRSN, is thus part of IRSN's drive to improve transparency and share knowledge with society at large.

Based on real-life experience, Grégory Rolina's study highlights a very special aspect of the assessment activities performed by IRSN: the continuous, in-depth technical dialogue between IRSN experts, ASN inspectors and facility licensees (primarily EDF, CEA and Areva). The author underlines the positive aspect of cultivating close relations with licensees: as experts gain knowledge of daily operating conditions at the facilities, they are able to incorporate real-life operating conditions into their reasoning, thereby improving the accuracy of their conclusions. When experts are able to convince the licensee, through in-depth technical discussions, that their recommendations are justified, it is reasonable to believe that compliance will be achieved more effectively than if the licensee only takes action to comply with official rulings.

Grégory Rolina also explains the risks involved in this approach. He calls attention to certain negative effects that could lend the licensee a significant power of persuasion. But the collaborative approach to IRSN assessments and the fact that they are based on the results of the Institute's own research, nonetheless serve to preserve its independence of judgement, for, within the context of technical dialogue, the licensee must convince the entire expert team involved. Some may be surprised to find the licensee participating in the assessment process, and the public may struggle to see how the expert can remain independent under these conditions. But instead of seeing this as a threat to the expert's independence, is it not preferable to emphasize that in this way, nuclear safety assessment adheres to one of the fundamental values of any scientific activity, which jurists refer to as an "adversarial procedure", something to be desired in any democratic society? Are we not in the sphere of illusion when we consider that an expert is independent from the licensee simply because there are no ties between them? Wouldn't a complete separation be detrimental to achieving an assessment that is scientifically relevant and ethically respectful of the licensee's word, as the primary actor accountable for safety in its facilities? These are some of the sensitive, troubling and particularly contemporary issues raised by the author, who avoids simplistic conclusions and provides the reader with tools that can be used to form his or her own opinion.

Furthermore, by seeking to define what constitutes an effective safety assessment in the processes implemented by IRSN, Grégory Rolina shows that successful "technical dialogue" requires skills that go far beyond the academic knowledge of those disciplines related to nuclear safety. The study clearly identifies the three key areas of IRSN expertise that contribute to the advancement of safety: building on past experience and ground-breaking research results to collectively establish recommendations conducive to taking concrete action in real-life situations, while also improving safety; generating new knowledge and lines of research based on an evaluation process; and finally, achieving the consent of peers and the safety authority through sound reasoning. Acquiring these skills is the baseline for training and professional development of each and every one of IRSN's experts.

Lastly, I would like to mention Grégory Rolina's particular focus on a specialized area of assessment, namely, the human and organizational factors involved in nuclear and radiological risks. Today, the vital importance of this field is widely acknowledged. Men and women play a decisive role in safety in high-risk industrial facilities, making human and organizational factors essential in any safety assessment. But how should they be evaluated? As this book demonstrates, scientific knowledge in this area is not yet firm enough to produce a sufficiently reliable frame of reference for assessments. A major challenge for experts in ergonomics, management and social sciences is therefore to gradually consolidate validated theoretical principles. The frame of reference alone, however, will never be enough: the author quite rightly emphasizes that an assessment of human and organizational factors must take full account of contextual and local knowledge of the human activities in question. The basic assumptions underlying the frame of reference must therefore be challenged frequently in light of empirical data, and discussed on a case-by-case basis with the licensee. Field experience is vital to the assessment approach used by IRSN's human factor specialists, as demonstrated in the examples provided by the author.

For all these reasons, I welcome the publication of this unique and informative study, which I warmly recommend to specialists in industrial risks and concerned citizens seeking to understand the workings of scientific and technical assessment in this area; it constitutes a significant contribution to the enhancement of nuclear safety in France and around the world.

Fontenay-aux-Roses, 15 May 2009

Jacques Repussard

Introduction

Thanks to the major scientific discoveries of the twentieth century, a global nuclear energy market has been built around the use of fissile matter. A concern to restrict application of this energy to the civil sector and the need to control the consequences for human health, soon led the United States, followed by the international community, to regulate and control access to this market. Thus it was that, in 1957, the International Atomic Energy Agency (IAEA) came into being, with the aim of promoting peaceful use of nuclear energy and establishing nuclear safety standards. In a widely-recognized document[1], the IAEA set out ten fundamental nuclear safety principles.

The first principle states that the responsibility for the safety of facilities or activities that give rise to radiation risks must rest with the person or organization responsible for operating such facilities or performing such activities (the licensee). The second principle emphasizes the role of national governments; "an effective legal and governmental framework for safety, including an independent regulatory body, must be established and sustained."[2] "Governments and regulatory bodies thus have an important responsibility in establishing standards and the regulatory framework for protecting people and the environment against radiation risks. However, the prime responsibility for safety rests with the licensee."[3] These principles, which call for the existence of

---

[1] AIEA (2006). Fundamental Safety Principles: safety fundamentals. Vienne, IAEA: 19.

[2] Ibid., p. 7.

[3] Ibid., p. 8.

a nuclear safety external review system in every country operating nuclear facilities, have been upheld in France for many years[4].

The French nuclear safety external review system includes a party that is absent from the IAEA's principles. France's regulatory authority regularly calls on experts, grouped within an organization. It is this expert assessment activity, which does not figure in the key international principles and is poorly described in institutional presentations, which is the subject of this book. Its contribution should help clarify the questions asked by the protagonists in a sector that is undergoing a major "renaissance", as well as by representatives of civil society, the media and researchers/lecturers in management science and in the social sciences. The original data on which our analyses are based were collected through intervention research. These data are compared with theories on expert assessment and risk control taken from the scientific literature. The specific nature of the activity studied is one of the points that will be elaborated on in this work.

1. CHALLENGES AND ISSUES IN THE BOOK

The existence of an organization that produces nuclear safety assessments is not specific to France's review system; Technical Safety Organizations (TSOs) were the subject of a first international conference in 2007[5], which aimed to explain good management practices in this type of organization. All the participants confirmed the existence of a TSO in the nuclear safety external review system in their countries and stressed the importance of its tasks. It would nevertheless have been useful to ask whether the organizations referred to by the acronym "TSO" all performed the same activity. For this reason, many of the participants, including the author, regretted the lack of detailed examples that could have been used to answer this question, all the more relevant given the great variety found in the statutes of the different TSOs, which may, for example, be a department of the regulatory authority, as in the United States and Sweden, or an independent institute where assessment and research

---

[4] However, in a certain sense, the independence of the regulatory authority was not consolidated until 2006, when France's Nuclear Security and Transparency (TSN) Act was passed. Prior to adoption of this law, the regulatory authority came under the supervision of a number of different ministries and therefore did not have the status of an independent administrative authority (cf. Gélard, P. (2006). Rapport sur les autorités administratives indépendantes. Paris, Office parlementaire d'évaluation de la législation: 136).

[5] *The Challenges Faced by Technical and Scientific Support Organizations in Enhancing Nuclear Safety*, a conference attended by around sixty countries, was organized by the IAEA and took place from 23 to 27 April 2007 in Aix-en-Provence (France).

activities are carried out, as in France, Germany and Belgium[6]. Other variables, such as the number of licensees, seem to have a significant impact on how the external review system works[7], and therefore probably on assessments.

The speakers at this international conference were not the only ones to point out the lack of concrete information on how nuclear safety assessments are produced. The lack of empirical research on the process by which external review systems assess industrial risks has been stressed by many representatives of academia, such as Mathilde Bourrier (2007) and Pierre-Benoît Joly (2007) in France. On the other side of the Atlantic, those representing the trend in research on high-reliability organizations say that "Nuclear utility/NRC[8] have received little attention at the microlevel. Most studies of the NRC have focused on macro issues such as the politics of regulation and regulatory reform rather than on case studies of the working relationships between plants and NRC inspectors."[9] According to these researchers, undertaking empirical investigations into how assessment works within external industrial risk control systems and how it affects facility safety would be a worthwhile cause. Examining the interactions between regulatory organizations and licensees is one of the possible extensions of Benoît Journé's thesis on safety management by operating teams at nuclear power plants (1999).

Among the various scientific disciplines upon which expert assessors in the external review system rely, human factors – often found in conjunction with organizational factors – is a subject likely to be of particular interest to researchers in management. Within French nuclear safety institutions, human factors emerged shortly after the accident at the US nuclear power plant at *Three Mile Island* (TMI) in 1979. Many analyses of this accident emphasized the lack of attention in facility design policy given to the conditions in which operating teams work. It is mainly to fill this gap that experts in ergonomics and the social sciences have been recruited to work with nuclear safety experts. The name of the discipline masks an ambitious project: to identify, anticipate and

---

[6] Le_Déaut, J.-Y. (1998). Rapport sur le système français de radioprotection, de contrôle et de sécurité nucléaire: la longue marche vers l'indépendance et la transparence. Such diversity may also explain the lack of international standards relative to TSOs, an eleventh fundamental nuclear safety principle. It may also reflect the existence of different models of assessment activity for radiological risk control.

[7] This, in particular, is shown by Rees, J. V. (1994). *Hostages of each other: the transformation of nuclear safety since Three Mile Island.* Chicago, The University of Chicago Press.

[8] *Nuclear Regulation Commission,* the US regulator for civil nuclear safety.

[9] La_Porte, T. R. and C. W. Thomas (1995). "Regulatory compliance and the ethos of quality enhancement: surprises in nuclear power plant operations." *Journal of public administration research and theory* **5**(1): 109–137, p. 114.

predict, in the context of organized technological processes, events involving human beings – in other words, nearly every type of event liable to lead to an undesirable accident – not all of which can be known in advance. Today, human factors experts are found in many fields, such as chemistry, petrochemicals, rail and air transportation. Their presence is justified by human involvement in technological processes and by the frequently inestimable cost of the consequences of an accident[10]. More than twenty years after Three Mile Island, the IAEA document mentioned above stipulates that "to prevent human and organizational error, human factors must be taken into account ...."

It appears that this field of specialization has been successfully incorporated in industrial sectors characterized by specific technological processes. It would be rash, however, to attribute this success to recommendations made by human factors experts. On the one hand, there appear to be gaps in the knowledge on which such recommendations are based – the links of cause and effect between human factors and safety seem tenuous; on the other hand, the effects of these recommendations are difficult to assess – they cannot simply be identified with reference to centralized decision-making. These two conjectures are developed in more depth in this study. How do assessments conducted by human factors experts impact safety at nuclear facilities? What type of knowledge are these assessments based on? These are the questions that we will attempt to clarify.

For this purpose, we will explain and analyze the process engaged in producing[11] human factors assessments, while seeking to identify the basic operations involved, placing a particular focus on the experts' interactions with other parties in the external review system, the representatives of the regulatory authority and the licensees. We will also reveal the various outcomes resulting from these operations, in particular, their effects on safety at the nuclear facilities in question. Secondly, we will define the types of effectiveness that characterize the assessment activity analyzed, then propose effectiveness criteria for each type, and finally, identify areas for improvement.

---

[10] It should be mentioned that human factors as a discipline is not necessarily exercised within the framework of an external risk control system; experts in human factors work at the companies in question.

[11] While this concept is traditionally used in the case of industrial goods, many researchers in the field of management have suggested applying it to services, stressing the importance of organizational procedures and management tools in the service sector (Bancel-Charensol, L. and M. Jougleux (1997). "Un modèle d'analyse des systèmes de production dans les services." *Revue française de gestion* (113): 71–81).

2. THE RESEARCH METHOD

The empirical data on which the analysis is based were obtained through intervention research, a methodology used mainly by researchers at the Centre de Gestion Scientifique (Scientific Management Centre) of the ParisTech department of the Ecole des Mines. Historically, this approach was primarily implemented to make up for a certain inadequacy in the use of interviews when attempting to comprehend how organizations really function. "When you interview people ... and ask them to describe their behavior, their answers are not focused on the behavior in question, they talk about something completely different, or about official behavior, but, in any case, not about behavior in the sense of saying, 'you know, in any case, I have to stick to an objective of such-and-such volume' .... Obviously, they never talk about that. So you have to do much more meticulous groundwork to try and find out what it is that makes them 'stick to the job'."[12] This groundwork generally takes place through close collaboration between researchers and the representatives of an organization, who, in making a request, often initiate the project. Partnerships can stretch over quite long periods (several years), and require coordination bodies to ensure that they proceed smoothly.

To gather pertinent data about an unfamiliar activity performed within institutions that are presumably opaque, and to avoid conventional discourse on effectiveness, intervention research thus seems to be a suitable methodology. It was carried out in collaboration with a team of human factors specialists from the French TSO, the Institute for Radiological Protection and Nuclear Safety (IRSN). These specialists work together in IRSN's Department for the Study of Human Factors (SEFH), under the Reactor Safety Division (DSR).

The partnership lasted for almost four years, including an initial period devoted to an assignment carried out as part of a post-graduate degree study.[13] The intervention research was divided into two stages, defined according to the successive requests made by the head of the SEFH.

The aim of my first assignment was to review human factors assessments produced since 1989. The head of the SEFH, who was my main contact at the time, had just taken over as department head and wanted a clearer vision of the department's past. By searching the archives, reviewing the scientific literature

---

[12] Comments by Jean-Claude Moisdon interviewed by Patrick Fridenson (Fridenson, P. (1994). "Jalons pour une histoire du centre de gestion scientifique de l'Ecole des mines de Paris. Entretiens avec Jean-Claude Moisdon et Claude Riveline." *Entreprises et Histoire* **7**: 19–35).

[13] Post-graduate degree in scientific management methods at the *Université Paris Dauphine*, which subsequently became a post-graduate degree in modeling, optimization, decision-making and organization.

and holding around thirty interviews with members of the SEFH, experienced specialists in human factors and representatives of IRSN, I was able to identify the key developments in the SEFH since it was set up following the TMI accident. My dissertation viva marked the end of the first stage of this assignment. It continued with the production of various statistics on the department's activity (number of cases per year, issues investigated, types of licensee, types of facility, etc.), based on the characterization of all the cases examined since 1989 using keywords.

At the same time, the head of the SEFH, who had suggested that I should study for my PhD thesis within his department, was thinking about the content of future research, which would soon need to broach the question: "What constitutes an effective assessment?" The question was phrased in this way following discussions between the head of the SEFH, my thesis supervisor and myself. Whereas the IRSN representative was interested in a study on his department's performance, my thesis supervisor's idea was that the study should approach the subject in terms of management science. Beyond a sociological analysis of assessment, the idea was to adopt a manager's viewpoint, by trying to establish a relationship between the way in which an assessment is constructed and the degree to which it is effective (i.e. the extent to which it successfully achieves the assessment objectives), which needed to be demonstrated.

The doctoral student, the thesis supervisor and the head of the SEFH also felt that it was essential to gather empirical data, which would help describe the assessment process in its entirety. In fact, the first assignment had revealed the limits of only looking at assessment reports. For the thesis, it was necessary to understand and analyze the production system that results in these reports in order to analyze the attributes of effective assessment. For this purpose, we needed to gather data on the assessment experts' practices, something to which interviews, however numerous, would not have given us access. So we decided to accompany the assessment experts on the job. A number of cases were selected, in agreement with the team of human factors specialists who accepted my presence alongside them as they worked.

During my thesis research period, a steering committee made up of expert assessors and researchers met ten times to check on progress. A great deal was learned from these meetings; this "management body"[14] for the intervention

---

[14] Girin, J. (1987). L'objectivation des données subjectives. Eléments pour une théorie du dispositif dans la recherche interactive. *Qualité et fiabilité des informations à usage scientifique en gestion.* Paris, FNEGE.

research provided a regular opportunity to discuss results or refocus the investigation.

Like the PhD thesis which preceded it, this book devotes a great deal of space to the empirical research carried out. The data gathered are original research material[15], which will be compared with models and theories regarding expert assessment and risk control.

3. MODELS OF ASSESSMENT ACTIVITY, FORMS OF CONTROL AND CAPTURE THEORY

The nuclear safety expert assessment that we set out to study focuses solely on risk control in civil nuclear facilities. Therefore, of the various studies that have explained models of activity, we have selected those devoted to assessment as well as those involving risk control. Although the latter term is employed by researchers in management, the term "regulation" is preferred by economists and political scientists. Some of them have developed a theory, known as "capture theory", which will be used here, since the French system of radiation risk control has often been criticized for being opaque, and thus potentially subject to the phenomena evoked by the notion of "capture".

3.1. Models of Assessment Activity

Within the community of researchers which has taken particular interest in scientific assessment processes since the 1980s, two types of expert assessment are commonly distinguished, according to the type of institution using it: judicial assessment and scientific assessment for political purposes. Judicial institutions began using experts early on; thus, "Roman law accepts the use of experts when know-how is required: the expert is called on to measure and assess."[16] This long history partly explains the predominance of the judicial model in mentalities. To explain a "spontaneous" representation of assessment activity, i.e. a canonical model, we will thus draw partly on the principles of judicial expert assessment.

---

[15] To our knowledge, no risk assessment activity has been as thoroughly investigated and reported on in this way.

[16] Leclerc, O. (2005). *Le juge et l'expert. Contribution à l'étude des rapports entre le droit et la science.* Paris, LGDG, p. 27.

The canonical model

Before examining the veracity of judicial expert assessment, Olivier Leclerc (2005) set out some of the principles on which it is based. These principles can be associated with a traditional representation of assessment. "Expert assessment can be described as a measure whereby scientific knowledge is put at the disposal of a judge in charge of settling a dispute. The expert assessment thus fulfils a decision-aid function: the expert provides the judge with factual elements that the latter integrates into the decision process. This distribution of functions came into being under old law and today structures judicial expert assessment in French law. The expert has a specific competence which escapes the judge: he must place it at the latter's disposal and may not, on any account, interfere in the actual function of judging."[17] This distribution of roles is justified by the logical bases of legal reasoning: "A judgement can be described as obeying the logical structure of syllogism. The applicable legal rule forms the major premise; the minor premise covers the facts that must be established and which constitute one or more conditions necessary to apply the legal rule. The conclusion of the syllogism is then drawn: the effects of the legal rule apply – or do not apply – to the facts established."[18] The syllogistic structure then guarantees the independence of the judge and the expert, i.e. the theoretical confinement of the expert to the field of fact; "the judge controls the entire syllogistic operation whereas the expert is only involved in the minor premise."[19] These first elements determining the functions of judge and expert and their relationship can be summed up by the first three proposals in Sidebar 1.

Part of Olivier Leclerc's thesis is devoted to the institutional mechanisms of judicial expert assessment. The French legal system has the particular feature of pre-selecting experts, who will be appointed by judges; the lists of court-accredited experts are the "key tool in expert selection in French law."[20] "The expert's competence is guaranteed by him/her being included on the lists and by his/her knowledge being certified before the lawsuit even takes place. When an expert is involved in a court case, the matter of his/her competence has already been addressed."[21] Two consequences of these statements are of particular interest to us:

---

[17] Ibid., p. 67.

[18] Ibid., p. 80.

[19] Ibid., p. 81.

[20] Ibid., p. 200.

[21] Ibid., p. 253.

1. French law considers the expert's knowledge to be independent of the expert himself; any person identified on a list of experts can be replaced by another person from the same list;

2. French law considers that the competence justifying expert status is acquired outside the expert assessment process.

They form proposals E4 and E5, which can also be associated with a traditional and spontaneous representation of expert assessment (cf. Sidebar 1).

Other properties complete our canonical model. It is accepted, in connection with proposals E4 and E5, that an expert has a body of pre-established knowledge external to the expert assessment process. To answer the question put to him, the expert must gather data. And to do so, he enjoys considerable freedom of action. By classic deductive reasoning, he then uses the data gathered and answers the question posed. This marks the end of the assessment process, which emerges as a relatively solitary process; if several experts are appointed, they are not supposed to cooperate; each one submits his/her recommendations, which result from a series of individual actions. This traditional representation of the expert assessment process, which relies on a positivist conception of science, allows us to set out proposals E6, E7, E8 and E9 of Sidebar 1.

The nine proposals made form the canonical model of expert assessment which we will use in the rest of this document. Some characteristics of this model do not firmly withstand the test of reality, which is namely what Olivier Leclerc's work shows.

> **E1.** A judge refers a matter to the expert to complete the knowledge required to pass judgment.
> **E2.** The expert's role is confined to facts.
> **E3.** The expert does not make any legal appraisal.
> **E4.** The expert's knowledge transcends the expert himself.
> **E5.** The expert's learning processes are independent of the assessment process.
> **E6.** The knowledge consists of a series of pre-established statements and know-how combined with a set of specific data gathered by the expert.
> **E7.** The expert is free to decide on the procedures he applies to answer the question raised.
> **E8.** The expert assessment is an individual process.
> **E9.** The expert's involvement is limited and does not continue after the judgment.

Sidebar 1: Nine proposals characterizing the canonical model of expert assessment.

Criticism of the canonical model in science of law

In his analyses, Olivier Leclerc mainly challenges proposals E2 and E3. Drawing on considerations from philosophy of law and knowledge, he asserts that an expert inevitably interprets the facts in a legal light; "A fact is not an 'existing circumstance' that must simply be examined in the light of the legal rule. Judgement of fact is based on a double intellectual operation:firstly, apprehension of the fact is a 'carefully considered undertaking' based on a selection of 'pertinent' factors (having regard to the legal rule to be applied) out of the given facts and, secondly, the 'hard material facts' are themselves interpreted by means of intermediary categories, taken from law and, more generally, from current language."[22] This is particularly the case in certain areas, in which expert assessment would "get to the truth of things, the judge then having only to pronounce the judgement which can be automatically deduced from the finding. That being the case, depriving a party of the benefit of an expert assessment amounts to depriving it of recourse to the judge, which the law expressly forbids."[23] Leclerc refers to the example of DNA: "scientific knowledge of the DNA profile offers such precise knowledge of the family relationship that it cannot be ignored by the judge: a judge must order an expert assessment since it gives him almost certain knowledge of the filiation."[24]

---

[22] Ibid., p. 96.

[23] Ibid., p. 117.

[24] Ibid., p. 115.

Thus, the expert's normative function must be acknowledged and the syllogistic structure of the expert-judge relationship revised. Particularly as Leclerc also points out that the judge cannot be confined to the strictly legal field: "a judgement does not merely consist in pronouncing the meaning of a norm independently of the pragmatic context to which it applies."[25]

Olivier Leclerc's criticism thus calls for re-conceiving the expert-judge relationship and the role these two functions play in judicial institutions[26]. It particularly invalidates proposals E2 and E3 of the canonical model which other research into scientific assessment calls for reviewing in greater depth.

An alternative model

The alternative model that follows is inspired by a review paper by Pierre-Benoît Joly (2005). It is the result of research into scientific assessment for political purposes carried out by sociologists, political scientists and jurists from the 1980s onwards, a period in which the occurrence of health and industrial crises greatly undermined people's confidence in scientific assessment. The critical view that society took of expert knowledge met that of researchers in human sciences, and expert assessment could no longer rely on principles which give science excessive authority. One alternative consists in governing expert assessment by procedures.

In the procedural model, "when involved as an expert in a complex field, a scientist always functions, whether consciously or otherwise, as the advocate of a certain cause."[27] Consequently, contrary to what proposal E4 of the canonical model upheld, experts in a given specialized area are no longer replaceable. To make up for the shortcomings of scientific knowledge and to achieve "reasonable knowledge as objectively reasoned as possible"[28], several researchers recommended providing a framework of procedures for scientific

---

[25] Ibid., p. 190.

[26] Similarly, in the field of scientific assessment, as Yannick Barthe and Claude Gilbert state, a lot of research has criticized the supposed political neutrality of science at work in the context of expert assessments. cf. Barthe, Y. and C. Gilbert (2005). Impuretés et compromis de l'expertise, une difficile reconnaissance. *Le recours aux experts. Raisons et usages politiques.* L. Dumoulin, S. L. Branche, C. Robert and P. Warin. Grenoble, Presses universitaires de Grenoble: 43–62 (479).

[27] Roqueplo, P. (1997). *Entre savoir et décision, l'expertise scientifique.* Paris, INRA Editions, p. 46.

[28] Ibid., p. 67.

assessment, particularly to ensure that all theories are put forward[29]. To respond to attacks on scientific assessment, these procedures must also guarantee that it is independent and transparent. The introduction of such procedures, which restrict the expert's freedom in choosing his methods and establish the collaborative nature of expert assessment, challenges proposals E7 and E8 of the canonical model. The model that ensues, described as procedural, is characterized by the three proposals in Sidebar 2.

E'1. Procedures guarantee the expert's independence.
E'2. Procedures guarantee transparency of the expert assessment process.
E'3. With procedures, open debates can be organized, thus guaranteeing that different theories are expressed.

Sidebar 2: Three proposals characterizing the procedural model of assessment.

These models of assessment activity will help us qualify human factors assessment. Since the assessment process is incorporated into a risk control system, we use theories from management science relative to control.

3.2. Forms of Control

From the wealth of literature on control, we content ourselves, in this brief introduction, with the seminal article by William Ouchi (1979), which brought to light several forms of control in force in organizations. The breakdown that Ouchi proposes is very general and can be applied to situations of risk control in industrial facilities.

Ouchi distinguishes three ideal types of mechanism, that of the market (control by prices), that of a bureaucracy (control by rules) and that of a clan (control by traditions). "The ability to measure either output or behavior which is relevant to the desired performance is critical to the 'rational' application of

---

[29] Hermitte, M.-A. (1997). "L'expertise scientifique à finalité politique. Réflexions sur l'organisation et la responsabilité des experts." *Justices*(8): 79–103, Roqueplo, P. (1997). *Entre savoir et décision, l'expertise scientifique*. Paris, INRA Editions, Godard, O. (2003). Comment organiser l'expertise scientifique sous l'égide du principe de précaution?, PREG CECO Laboratoire d'économétrie: 18, Noiville, C. (2003). *Du bon gouvernement des risques*. Paris, Presses universitaires de France. We should further mention that in conclusion to a whole series of intervention research studies on decision-aid models in public transport policies, researchers at the scientific management centre already recommended a set of procedures to provide a frame of reference for quantitative studies. They particularly insisted on the adversarial principle. cf. GRETU (1980). *Une étude économique a montré ... Mythes et réalités sur les études de transports*. Paris, Editions Cujas.

market and bureaucratic forms of control."[30] If the output cannot be measured, and the rules of behavior cannot be defined, the only possible form of control is the clan mechanism. Ouchi illustrates this situation with the example of a research laboratory: "The organization relies heavily on ritualized, ceremonial forms of control. These include the recruitment of only a selected few individuals, each of whom has been through a schooling and professionalization process which has taught him or her to internalize the desired values [characteristic of the laboratory] and to revere the appropriate ceremonies."[31] The resulting forms of control are given in Table 1.

		Knowledge of the rules by which results are achieved	
		Perfect	Imperfect
Ability to measure results	High	*Control by result or compliance with the rules*	*Control by result*
	Low	*Control by compliance with the rules*	*Clan control (control of levels of competence)*

Table 1: Forms of control (adapted from Ouchi 1979).

To apply Ouchi's principles to radiation risk control, the following two questions must first be answered: can we measure the results of a safety production process and do we know the rules by which these results can be achieved? The definitions of safety proposed by Karl Weick and Erik Hollnagel leave room for doubt: safety is a "dynamic non-event"[32], "... the sum of the accidents that did not occur"[33].

If we follow Ouchi, we should therefore expect to observe clan forms of control. However, empirical works and industrial testimonials would appear to identify others. Researchers who studied high-risk organizations, namely in the nuclear, chemical, aeronautical, rail and oil industries, all noted the significant number of procedures[34]. These sectors appear to be dominated by a safety

---

[30] Ouchi, W. G. (1979). "A conceptual framework for the design of organizational control mechanisms." *Management science* **25**(9): 833–848, p. 843.

[31] Ibid., p. 844.

[32] Weick, K. (1987). "Organizational culture as a source of high reliability." *California management review* **29**(2): 112–127, Weick, K. and K. Sutcliffe (2001). *Managing the unexpected.* San Fransisco, Jossey-Bass.

[33] Hollnagel, E. (2006). Resilience – the challenge of the unstable. *Resilience engineering.* E. Hollnagel, D. D. Woods and N. Leveson. Hampshire, Ashgate: 9–17.

[34] At a conference, a representative of EDF stated that a nuclear power plant was governed by 30,000 procedures!

strategy based on anticipating risky situations[35]. In the nuclear field, this is reflected in extensive use of the notions of "defense in depth" and "barrier", and the presence of bureaucratic-type control (compliance with procedures, checking the existence of barriers, demanding extra barriers). Furthermore, licensees clearly advertise the importance they attach to feedback, known as "operating experience feedback" or "OEF", and implemented by reporting procedures in the event of incidents, by triggering analyses and by processing statistics. These procedures are particularly used to count the number of "safety significant" events and to establish indicators for a form of control by results.

Without challenging the validity of Ouchi's theory, the identification of these various forms of control reflects the existence of several conceptions of safety. When taken as a number of incidents, safety can be controlled by the result; when taken as a set of mechanisms that enable prevention of an accident, safety control is bureaucratic (compliance with a set of procedures to ensure safety systems exist); when taken as a dynamic non-event, risk control should be of the clan type. Our empirical data should therefore enable us to illustrate the conception(s) of safety adopted by human factors experts.

The breakdown proposed by Ouchi is not specific to risk control exercised by a regulator external to the company. Other research work has been done on this configuration and some highlights a particular form of relationship between the regulator and the regulated.

3.3. The Capture Theory

American economists and political scientists[36] have given particular focus to the study of situations involving risk control by public authorities. Some of their analyses are based on the notion of capture, greatly echoed in public debates. "The capture theory, a tenet of academic political science, was popularized in the late 1960s and early 1970s, by a steady stream of exposés of federal agencies by[37] Ralph Nader's 'Raiders'. Regulatory officials were pictured as

---

[35] Wildavsky, A. (1988). *Searching for safety.* New Brunswick, Transaction publishers, Journé, B. (1999). Les organisations complexes à risques: gérer la sûreté par les ressources. Etude de situations de conduite de centrales nucléaires. *Sciences de l'homme et de la société. Spécialité Gestion.* Paris, Ecole Polytechnique. **Thèse de doctorat**: 434, Hood, C. and D. K. C. Jones, Eds. (1996). *Accident and design. Contemporary debates in risk management.* Abingdon, University College London Press.

[36] Stigler, G. J. (1971). "The theory of economic regulation." *Bell journal of economics and management sciences* **21**: 3–21, Peltzman, S. (1980). "The growth of government." *Journal of law and economics* **23**: 209–287, Bardach, E. and R. A. Kagan (1982). *Going by the book. The problem of regulatory unreasonableness.* Philadelphia, Temple University Press.

[37] Ralph Nader was namely behind the creation of *Raiders,* in the late 1960s, a well-known consumer protection group in the US.

industry-oriented, as reluctant to jeopardize their postgovernment careers by being too tough, or as gradually co-opted by informal contact with representatives of regulated firms."[38] When the regulator is captured, he becomes the regulated's advocate; the regulation system is corrupt.

To avoid capture phenomena, several authors have recommended introducing systems: "Breaking up relationships between regulators and regulated (by periodic staff rotation and revolving door employment restrictions [for regulators]), expanding citizen groups' rights to participate, stipulating that agency decisions must be made in public session, providing appeal mechanisms for any aggrieved party, allowing private parties to seek judicial orders to force regulatory agencies to take action, opening agency records to reporters and others, providing federal reviews of state agencies' actions, forbidding advance notice of inspections, and centralizing the scheduling of inspection programs to remove decentralized discretion."[39]

While excessively close ties between regulators and regulated can lead to forms of capture that are ineffective, nearing corruption, some authors, including Ayres and Braithwaite (1992), have referred to efficient forms of capture, which can have positive effects on risk control. Using economic modeling based on the game theory, they demonstrated that in some cases, the optimal strategy was achieved when the two parties cooperated. Furthermore, La Porte and Thomas (1995) identified a positive form of capture by studying the relationships between a regulatory authority inspector (NRC) and the staff of a US nuclear power plant. The inspector was a resident, i.e. based at the plant. La Porte and Thomas noted considerable proximity between the inspector and the plant staff, which could have led them to confirm the capture of the regulator. Yet, they noted that this proximity, particularly via the inspector's involvement in the plant's ongoing improvement systems, prompted the licensee to set higher safety standards than those of the NRC.

Application of this theory is justified by the suspicions that weigh on the nuclear industry, its various agencies forming the lobby targeted by the invectives of civil society representatives. Our empirical data may also identify capture avoidance systems.

The different models of expert assessment activity, forms of control and the notion of capture thus constitute the theoretical tools we will use to interpret our data.

---

[38] Bardach, E. and R. A. Kagan (1982). *Going by the book. The problem of regulatory unreasonableness.* Philadelphia, Temple University Press, p. 44.

[39] Sparrow, M. K. (2000). *The regulatory craft.* Washington, The brookings institution, p. 37.

4. ARGUMENTS AND OUTLINE OF THE BOOK

While use of assessment activity models and forms of control requires no further justification, points of history regarding French nuclear safety institutions will underline the pertinence of referring to the capture theory. The historical perspective will also focus on the main principles of nuclear safety, and on the emergence and inclusion of the human factors specialty in the French expert assessment institute. This presentation will provide important insights into one of the central points of our work: the nature of knowledge in human factors assessment. Particular emphasis will be placed on the importance of a conception of safety based on the notions of barrier and defense in depth; on the extension of objects assessed by human factors experts (from human factors to human and organizational factors); and on the need – and the difficulty – for human factors specialists to obtain empirical data, i.e. resulting from an investigation requiring access to nuclear facilities. The historical overview of human factors assessment will conclude with a presentation of the current Department for the Study of Human Factors and how its expert assessment work is organized. This inventory will provide an initial representation of the human factors assessment activity. [Part 1]

Three standard cases, representative of the aforementioned department's expert assessment, were selected as the empirical material for the research. By closely following the experts in action, analyzing numerous documents and conducting additional interviews, data was obtained to report on the various expert assessment processes in their entirety. These detailed reports reveal a human factors assessment activity which is not in line with the models presented above. The canonical model of expert assessment is particularly inappropriate to characterize the cases studied and the procedural model is also insufficient; it ignores key operations of the assessment production system revealed. The material particularly shows how the expert's learning is inseparable from assessment processes. Furthermore, the various forms of risk control are insufficient to account for the experts' investigation, which generally includes a step of exploring incident scenarios or examining the organizations assessed. Lastly, the presence of negotiations with the licensee throughout the assessment process poses a threat to the expert's freedom; our observations do however, refute any capture of the expert by the licensee. [Part 2]

The main result of assessments conducted by human factors experts is a list of findings and recommendations which, when first analyzed, reveal gaps in the knowledge on which they are based. To improve that knowledge, specialists may build chains of events involving human (and organizational) factors that could lead to a risk; by examining the links between human factors and safety, they help improve what we have called the "cognitive effectiveness of human factors assessment". However, compliance with a set of institutional requirements and the need to achieve certain vital objectives (recommend and "align" points of view) can prompt experts to focus primarily on what we call the "rhetorical effectiveness of expert assessment", sometimes to the detriment of improving knowledge. To examine the effectiveness of expert assessment, the impacts of human factors assessment on nuclear facilities must naturally be studied. Together, these effects can be tied to a third dimension of the effectiveness of assessment, referred to as operational. To attempt to evaluate it, expert assessment must be regarded both as one element in a series of assessments and as a process of interactions between regulators and those they regulate. Following this analysis, we summarize all the skills that the expert must have to achieve effective assessment, in the rhetorical, cognitive and operational sense. [Part 3]

By distinguishing and characterising three types of action performed by human factors experts, this study focussing on the effectiveness of their work enables us to enhance the representations of expert assessment that result from the models we have presented. [General conclusion]

Part One.
Technical Dialogue and Human Factors:
a Historical Perspective

"Each society has its regime of truth, its 'general politics' of truth: that is, the types of discourse which it accepts and makes function as true; the mechanisms and instances which enable one to distinguish true and false statements, the means by which each is sanctioned; the techniques and procedures accorded value in the acquisition of truth; the status of those who are charged with saying what counts as true"
Michel Foucault

Human factors are a specialized area of nuclear safety that did not emerge in France until the early 1980s. In order to better grasp the emergence of nuclear safety issues and how they are addressed by IRSN's experts, we have set them against the historical backdrop of France's nuclear safety institutions and policies. Taking a historical perspective, which is suggested by many researchers in management, helps in understanding current practices.

In the years following World War II, national choices regarding nuclear energy policies culminated in the creation, in the 1970s, of a safety triad: that of the licensee, the authority and the expert. This triad, dubbed "French cooking" by the Americans and often criticized for its opacity, is characterized by specifically French traits (licensees have public enterprise status, are few in number and are staffed primarily by graduates of France's elite schools of engineering). The year 1979, marked by the accident at *Three Mile Island*, was a watershed in the history of the world's nuclear industry. In France, an ambitious action plan based on analyses of the accident was developed. The findings of all of these analyses emphasized that, during the design stage of the Three Mile Island nuclear power plant, scant notice was given to the working conditions of the plant's operators. Ergonomics experts were called in to implement corrective actions, marking the emergence of human factors in the nuclear safety triad.
[Chapter 1]

Within the safety assessment institution, human factors experts worked together in a laboratory set up with the main objective of building up a body of knowledge. The difficulties encountered by the laboratory's experts led to a crisis in 1988, when two-thirds of its staff left en masse. As counterintuitive as it may sound, it was necessary to change the laboratory's focus in order to resolve the crisis. Despite having amassed knowledge in their field, the human factors experts turned their attention more to assessments rather than studies. In addition, other issues entered the field of human factors. Management, safety culture and organizational factors, which were identified as elements that contributed to the Chernobyl disaster, became crucial to safety and had to be assessed by human factors experts. The Chernobyl disaster also marked the starting point of a movement to translate the values of independence and transparency within the safety triad into tangible action. It typified the recent changes that had affected the institutional context of the activity of human factors experts. Several types of human factors assessments were developed, with organizational conditions defined for each. These conditions might draw the assessment closer to the procedural model. [Chapter 2]

This is one of the postulates that we will state at the end of this historical and institutional presentation. [Conclusion to part one]

Chapter 1. The Emergence of Human Factors in Institutions of Technical Dialogue

The history of nuclear safety in France is characterized by a sort of "organizational internalization of risks", that is, a risk management process maintained within the organizations involved in nuclear power."[40] After World War II, nuclear safety and control policies, in large part imported, were developed within France's Atomic Energy Commission (CEA), the body responsible for promoting nuclear energy. The 1970s were marked by the implementation of an ambitious nuclear power program based on pressurized water technology imported from the United States and entrusted to France's state-owned electric utility Electricité de France (EDF). This led to the creation of an administrative authority in charge of controlling nuclear facilities. This authority drew on the expertise of the CEA, which retained its safety analysis capacity. The nuclear safety institutions thus established were ones of "technical dialogue" and "French cooking". The *Three Mile Island* accident in the United States in 1979 did not put a halt to the vast program to build nuclear power plants. One of the solutions used to cope with the trauma caused by the accident can be summed up in two words: human factors. In France, a team of human factors experts was formed within the CEA's assessment institute.[41]

---

[40] Vallet, B. (1985). *La constitution d'une expertise de sûreté nucléaire en France*. Situations d'expertise et socialisation des savoirs, Saint-Etienne, p. 14.

[41] The historical analysis presented herein is taken from the following three publications in particular: Cyrille Foasso's 2003 PhD thesis in history, a report submitted in 1984 to the administrative authority on nuclear safety by sociologist Bénédicte Vallet, and Anne-Sophie Millet's 1991 PhD thesis in law. A few accounts from former experts are also included. Much has been written about the Three Mile Island accident. We have drawn on analyses by French and American experts explaining the causes of the accident and its consequences on nuclear safety institutions and policies.

1. THE ADVENT OF NUCLEAR SAFETY IN FRANCE AT THE CEA

The French Atomic Energy Commission (CEA) was created in October 1945. The great ambitions General de Gaulle had for the scientific programs led by Frédéric Joliot-Curie, high commissioner of the newly created CEA, allowed for very broad freedom of action: "Placed under the authority and control of the president of the council of ministers, the new commission would benefit from a new status unique in France. Endowed with a legal personality, it was administratively and financially autonomous."[42]

At the time, nuclear safety was unheard of. The page was blank and everything had yet to be written. "Before 1960, safety didn't exist as such. It was completely integrated in projects and facilities promoted by people who were already practicing safety without knowing it."[43] No one was specifically appointed to formulate rules on safety or safety-related training. "The technicians themselves were responsible for ensuring safety during experiments, be it by designing systems or establishing and enforcing rules."[44] At the time, the CEA had the monopoly on scientific activities and was the sole source of know-how in France's nuclear technology industry. "The very first reactors were developed at the CEA without any requirement for external oversight."[45] It was entirely through their daily experiments and with a high degree of autonomy that the CEA's scientists began to apprehend safety. In the early 1960s, France caught up with other nuclear powers that had already "singled out" safety by forming groups of experts and thus "unbinding" it from the practices of physicists. A commission for safety at atomic facilities was set up at the CEA and policy principles, often imported, were adapted to France's nuclear industry, which was characterized in particular by very loose regulations. These choices formed the framework for "technical dialogue" between facility licensees and safety experts, the foundation of the regulation of French nuclear facilities.

---

[42] Goldschmidt, B. (1980). *Le complexe atomique. Histoire politique de l'énergie nucléaire.* Paris, Fayard, p. 137.

[43] Cogné, F. (1984). "Evolution de la sûreté nucléaire." *Revue générale nucléaire*(1): 18–32. François Cogné was a senior official of French nuclear safety institutions (director of the Institute for Nuclear Safety and Protection from 1985 to 1988; chairman of the Advisory Committee for Nuclear Reactors from 1985 to 2001) *(sources: Foasso, 2003)*.

[44] Foasso, C. (2003). Histoire de la sûreté de l'énergie nucléaire civile en France (1945–2000). Technique d'ingénieur, processus d'expertise, question de société. *Histoire moderne et contemporaine.* Lyon, Université Lumière – Lyon II. **Thèse de doctorat**: 698, p. 72.

[45] Vallet, B. (1984). La sûreté des réacteurs nucléaires en France: un cas de gestion des risques. Rapport au service central de sûreté des installations nucléaires, Ecole des mines de Paris – Centre de sociologie de l'innovation: 123, p. 22.

1.1. An Artisanal Approach to Safety

The absence of standards and external controls, typical of the time, is often cited as a catalyst for research. The following opinion published in Bénédicte Vallet's report is shared to a fairly large degree by members of the scientific community who participated in the pioneering work of the period: "With today's safety and security standards, we wouldn't be able to do now what we did back then. It would be impossible."[46]

This broad freedom of action allowed to physicists does not seem to have endangered the safety of workers and infrastructure. Historical research has not found that the development of French nuclear power experienced any major crises in the 1950s.[47]. Some attribute this success to the safe behavior of the CEA's scientists. "The way the people from the early days remember it, safety may not have been an objective or a specific area ..., but [it] was an integral part of all operations in the nuclear industry."[48] In his history of nuclear safety, Cyrille Foasso provides many examples that illustrate the concern of scientists to make safety systems a part of their experiments. One such example is the criticality event[49] at an experimental reactor: "Aware of the risk posed by the operation, they set up multiple means of control and protection. For example, they had a completely windowless building built to reduce the risks of leaks. The facility was equipped with control rods or shutdown rods[50] driven by a giant 'crossbow' so that they could be propelled at an extremely high speed."[51] The first activity report of the CEA (January 1' 1946–December 31, 1950) describes the operating conditions of France's first atomic reactor. Dubbed Zoé, it was commissioned on December 15, 1948. "During the first few months we operated it at only very low power (a few Watts) so that we could train the control room staff in conditions where operating errors would not have been serious and, after a sort of running-in period, shut down the reactor, dismantle it and perform a check without its radioactivity becoming too high." Reports Bénédicte Vallet: "Although [Zoé] wasn't equipped with a proper cooling system, the people in

---

[46] Ibid., pp. 79–80.

[47] Nonetheless, two incidents did occur: one in the G1 reactor in 1956 and another in the Alizé 1 reactor in 1959.

[48] Ibid., p. 23.

[49] Criticality corresponds to the start of the initial chain reaction.

[50] These rods are designed to slow down or stop the chain reaction.

[51] Foasso, 2003, p. 71.

charge ensured the safety of the reactor and its staff (staff training and studies of materials for future reactors) and introduced it in baby steps."[52]

Another factor is the good working relationships between scientists, which were conducive to the development of what today's risk management experts would refer to as "safety culture": "What was important in terms of safety was somewhat the mutual trust that the scientists shared with each other; a trust that the guys wouldn't try to 'get more from the reactor'."[53]

1.2. First Efforts Toward the Institutionalization of Safety

Thus were the early years of nuclear power "a period of artisanal research devoid of any specific safety rules apart from those that researchers, engineers and technicians voluntarily set for themselves."[54] Historians looking at the development of nuclear technology in the United States describe a similar situation during the early post-war years in America: "Because nuclear power was not far removed from the laboratory, only a handful of experts understood the complexities of nuclear power: those who designed reactors were also the leading authorities on their safety."[55] However, this situation did not last for long in the United States. Nobel laureate Enrico Fermi – who, with his team in Chicago, was the first to control a chain reaction in 1942 – was also "one of the first to recognize that the days of undifferentiated development and safety review were numbered."[56]

In 1946, well ahead of the other nuclear powers, the United States drafted nuclear safety legislation by passing the Atomic Energy Act (or McMahon Act). The Act established the United States Atomic Energy Commission (AEC). Just over a decade later, the Windscale fire of 1957 in Great Britain prompted the creation of agencies specifically charged with overseeing nuclear safety issues. In France, initial efforts to institutionalize safety began to take shape within the CEA itself in the early 1960s. Foasso identifies the following three reasons behind these efforts:

- *Nuclear power had entered a new era marked by the shift from experimental activities led exclusively by the CEA's engineers to industrial-scale production of plutonium and electricity involving other companies, notably EDF, whose*

---

[52] Vallet, 1984, p. 31.

[53] Ibid., pp. 79–80.

[54] Ibid., p. 76.

[55] Balogh, B. (1991). *Chain reaction: expert debate and public participation in American commercial nuclear power, 1945–1975.* New York, Cambridge University Press, p. 120.

[56] Ibid., p. 122.

first reactor reached criticality in 1962. There was thus a need for oversight and standardization in order to control these reactors, which were constantly gaining in capacity;

- *Furthermore, 1957 saw the creation of the International Atomic Energy Agency and Euratom, which was established by the Treaty of Rome. The multiplication of international standardization organizations whose rules could apply to all member countries was a threat to France's independence. The idea that a foreign or international authority could dictate to the best and brightest in French research and industry what standards had to be followed was insufferable;*

- *Lastly, Foasso lists a series of accidents and incidents that occurred in England, the United States, Yugoslavia, the Soviet Union, Canada and France in the late 1950s. These events prompted a tightening of reactor oversight and safety and the creation of the Atomic Plant Safety Commission (CSIA) within the CEA in 1960.*

1.3. The Development of Policies

So what did the CSIA do? To start off, it caught up. The safety of nuclear facilities was already the focus of discussions between American and British experts at the first Geneva Summit of 1955. The CSIA's members, and those of its subcommission on reactor safety in particular, took inspiration from their American and British counterparts. In the words of a key figure in the history of French nuclear safety, and who at the time was in charge of the aforementioned subcommission, "one visit to the United States was highly instructive: .... Some methods had been developed to prevent risks and mitigate their consequences if they ever occurred. These methods were applied to nuclear facilities in the following manner: at each stage of development of a facility (design, construction and operation), safety had to be demonstrated in a written report that contained a detailed description of the facility's condition, particularly regarding components important for safety and, more importantly, a 'maximum credible accident' study. The study's findings had to show that the maximum credible accident truly was the worst and would not cause any disruptions beyond the facility's site."[57]

---

[57] Bourgeois, J. (1992). La sûreté nucléaire. *L'aventure de l'atome.* P. M. d. l. Gorce, Flammarion.

Safety analysis report reviews

The approach adopted by the subcommission led by Jean Bourgeois, and which has been a prerequisite for the construction and commissioning of every facility since 1960, is based on American practice. According to Foasso, the relationship between the subcommission's members and nuclear facility operators was far from being one of pure oversight. For instance, during a session held to review safety issues during power ramp-up in the EDF 1 reactor in Chinon (December 6, 1962), the high commissioner emphasized "the close collaboration established between the CEA's experts and EDF's engineers", who "worked in symbiosis". Furthermore, as Vallet states (1984), it was EDF officials who requested the assistance of Jean Bourgeois' team on the matter. "When EDF built its first reactor, EDF 1, which was put into service in 1963, no department or agency at the time was able to organize the safety of facilities. According to several sources, EDF itself, aware of its lack of knowledge in the field, was behind the move to entrust the CEA with reviewing safety at EDF 1."[58]

Maximum credible accident analysis

The second American practice mentioned by Jean Bourgeois is the "maximum credible accident" analysis. The concept of MCA had been introduced by Clifford Beck, an expert with the AEC, in particular as an attempt to streamline the decision process related to the siting of future nuclear facilities. The following words, taken from a speech by Beck, properly set into context the major issue raised by facility safety: "If worst conceivable accidents are considered, no site except one removed from populated areas by hundreds of *miles* would offer sufficient protection. On the other hand, if safeguards are included in the facility design against all possible accidents having unacceptable consequences, then it could be argued that any site, however crowded, could be satisfactory ... assuming of course that the safeguards would not fail and that no potentially dangerous accidents had not been overlooked."[59] The assumption about the state of knowledge on possible accidents is remarkably optimistic and the confidence Beck proposed to place in the reliability of safeguards was no doubt undue. Neither did he mention the costs that the second extreme solution would incur if it were put into practice. MCA analysis, which could cause the

---

[58] p. 44. It should be noted that, at the time, the CEA was EDF's supplier of reactor fuel. "The CEA's engineers, who dealt with safety, were therefore perfectly aware of the problems posed by fuel and the resulting risks." (January 12, 2007 interview of a former safety expert).

[59] Taken from a speech by Clifford Beck of the AEC during the 1958 Geneva Summit, cited by Foasso, 2003.

most significant radioactive releases, was midway between the two extremes adopted by the Americans. Once an MCA was identified, the next step consisted in putting in place safety systems to avoid or even contain the undesirable event's consequences. William Keller and Mohammad Modarres suggest a concept of safety inferred from the MCA technique: "Safety was thus defined as the ability of reactors to withstand a fixed set of prescribed accident scenarios judged by AEC experts as the most significant. The implicit premise was that if the plant could withstand these design basis accidents it could withstand any other accident."[60]

For Foasso, MCA became a genuine bargaining chip between the US authority and US facility licensees. The relationship they share seems to be, more than is the case in France, a traditional one of oversight: "Under normal operating conditions, that is, in non-accident situations, the criterion was simple: standards on radioactive releases had to be met. In the event of an accident, the facility's design had to prevent the dose along the site's boundaries from exceeding a set value. Verification by the authority consisted in checking whether the consequences of a specific accident, the severest one whose likelihood was judged credible, did not exceed this standard. In other words, if the designers could prove that the amount of radiation released by their facility following an accident did not exceed this standard along the site's boundaries, authorization could be granted."[61] An account by a former top federal regulatory official offers a glimpse of the use of this concept: "The committee reviewed each reactor plan, trying to imagine an accident even worse than that conceived by the planner. If we could think of a plausible mishap worse than any discussed by the planner, his analysis of the potential dangers was considered inadequate."[62] Joseph Rees writes that over time this "simple procedure" became the foundation of the AEC's facility review process required in order to obtain a license to commence operation.[63]. One of the main reasons Foasso cites to explain this difference in the relationships between safety experts and facility licensees in France and the United States is that the American nuclear industry is "quite different from that [encountered] in France and Great Britain. There are

---

[60] Keller, W. and M. Modarres (2005). "A historical overview of probabilistic risk assessment development and its use in the nuclear power industry: a tribute to the late Professor Norman Carl Rasmussen." *Reliability engineering and System safety* (89): 271–285, p. 272.

[61] pp. 85–86.

[62] Mazuzan, G. T. and J. S. Walker (1985). *Controlling the atom*. Berkeley, University of California press, p. 61.

[63] Rees, J. V. (1994). *Hostages of each other: the transformation of nuclear safety since Three Mile Island*. Chicago, The University of Chicago Press, p. 30.

many electric utilities, several reactor builders, dozens of different models, hundreds of equipment suppliers. ... In addition to the number of organizations, the commercial aggressiveness of American companies [necessitates] the establishment of rules for dialogue."[64] The situation in France's nuclear industry was altogether different. Until 1969, the CEA was the sole source of French nuclear expertise, although EDF had also been operating nuclear reactors since 1962. French regulations were therefore enforced in a sector that was shared by two state-owned institutions. Technology was another differentiating factor between France and the United States Until the early 1970s, most reactors in France were of the gas-cooled graphite-moderated variety. Lastly, specialists, facility licensees, experts and inspectors in France were all graduates of the country's elite schools of engineering. They thus shared a common bond that helped to unite them whatever their position on the professional ladder.

The British specialist F.R. Farmer was highly critical of the concept of MCA. For Farmer, "focus should not be solely on a certain number of hypothetical disastrous accidents but also on other types of incidents with a higher probability of occurrence."[65] The underlying idea, which would be borne out by the *Three Mile Island* accident a few years later, was that defending oneself against the worst conceivable accident wouldn't necessarily make it possible to defend oneself from a somewhat less serious accident. Although French specialists would adopt the concept of MCA, they were aware of its limitations and thus combined it with another method, that of barriers.

The barrier method

The concept of barriers was raised in the Farmer's 1958 speech in Geneva, but not systematically as was the case in France from that period on.[66] Applying this method thus amounted to following Farmer's recommendation by considering different accident scenarios that could result in the release of radioactive substances in addition to an MCA. Then, as with MCA, barriers were "inserted" and their robustness was studied.[67]

The main policy principles adopted by France to ensure safety at its nuclear reactors were virtually established in the late 1960s. France had therefore caught up. However, as regards regulations, the country stood out as an exception.

---

[64] p. 90.

[65] Foasso, 2003, p. 230.

[66] Ibid., p. 126.

[67] As will be seen, these multiple layers of barriers are the central feature of defense in depth, a principle used in the design of nuclear power plants.

1.4. France's Half-Hearted Nuclear Safety Regulations

As Anne-Sophie Millet states in her thesis on French nuclear law (1991), while most nuclear countries quickly passed legislation governing the oversight of nuclear facilities, in 1963 France's government promulgated a simple decree distinguished by its weak tenor. Bénédicte Vallet lists several reasons that led to the promulgation of this decree on basic nuclear facilities[68]. Three are listed below:

- *The main reason was to put an end to the special bylaws of the CEA, which was exempt from any kind of external control;*

- *Like all other "facilities presenting a threat to health and safety or other hazards", the nuclear facilities that EDF had begun to operate were subject to the French law of December 19, 1917. Oversight of these facilities was thus placed in the hands of the departmental prefectures, which lacked the necessary skills;*

- *There was also the obligation to comply with the Treaty of Rome (Euratom), which required the creation of a system of official reports and authorizations that were necessary in order to begin activities involving radiation risks.*

The creation of every basic nuclear facility therefore had to be authorized in a decree signed by the French prime minister. "All draft decrees on the creation or substantial modification of basic nuclear facilities had to be approved beforehand by an Interministerial Commission on Basic Nuclear Facilities (CIINB), which was established by the decree of December 11, 1963."[69] For Millet, this step was crucial: "Safety expanded beyond the confines of public institutions and entered the reach of French government."[70] However, as Foasso points out, "the basic elements of safety were ultimately absent from the decree, since it did not set out the procedure on how experts were to conduct facility safety reviews. And yet this procedure made up the bulk of the construction permit review process." Likewise, Millet writes about the limited scope of the decree of 1963, stating that it "made no provisions for safety. It did not specify who could or had to step in

---

[68] Basic nuclear facilities are defined by article 2 of French decree No. 63-1228 of December 11, 1963 as nuclear reactors (with the exception of those forming part of a means of transport), particle accelerators, facilities used for the preparation, production or processing of radioactive substances, facilities designed for the storage, warehousing or use of radioactive substances, including wastes.

[69] Vallet, 1984, p. 47.

[70] Millet, A.-S. (1991). L'invention d'un système juridique: Nucléaire et Droit. *Droit*. Nice, Université de Nice-Sophia Antipolis. **Thèse de doctorat**: 625, p. 95.

between a facility's creation and its operation."[71] With no formal safety review procedure, the CEA would in practice continue to fill the void for a number of years.

1.5. Conclusion: the Birth of French Cooking

This period in the development of France's nuclear program was thus devoted to the establishment of policies that would henceforth characterize the French approach. Although the principles retained were often imported from the United States, the situation in each country was very different.

The US nuclear program, characterized by the necessity for safety experts working for an administrative authority to regulate the activity of a large number of parties, contributed to the creation of the concept of the maximum credible accident (MCA). MCA was a relatively simple tool that enabled experts, within the context of a relationship of results-based oversight, to make quick decisions about whether or not to license nuclear facilities. Back then, Farmer feared that MCA would turn into a "tool of compliance"[72], structuring the parties' behaviors around a scaled-down simplified issue that could run counter to nuclear safety: "The problem of assessing the relative safety of a reactor is complex and difficult. ... In such a situation there is need for vigilance to avoid being driven by pressures of administrative and legislative convenience to produce readily manipulated formulae or rules against which reactor safety is to be tested. Once this happens the efforts of designers and operators are directed to compliance with just those requirements, and further thought on the effect or their efforts on real and fundamental safety may well be neglected."[73]

In France, the relationship that was forming between facility officials and experts on the reactor safety subcommission was not really one of oversight. "The reviewers were not aloof, holier-than-thou judges handing down regulatory restrictions carved in stone. In fact, they were in constant contact with the people in nuclear facilities and thus understood their realities. Also, thanks to sustained analysis activities they were able to personally contribute to finding solutions to safety issues."[74]. This is the "technical dialogue", or "French

---

[71] Ibid., p. 143.

[72] Moisdon, J.-C., Ed. (1997). *Du mode d'existence des outils de gestion*. Paris, Seli-Arlsan.

[73] Taken from a speech by Farmer at the April 1967 IAEA Symposium in Vienna, cited by Foasso, 2003, pp. 230–231.

[74] Taken from an article by F. de Vathaire "La sûreté des réacteurs: réalisations et tendances actuelles", Energie nucléaire, vol. no. 7, pp. 421–427, cited by Foasso, 2003, p. 240.

cooking"[75], that regulated relations between the two parties. In such a system, characterized by the limited role of regulations, adopting a decision-making tool that would cut back on reality too drastically was not necessary. "The technical dialogue between experts was de facto institutionalized by the absence of nuclear regulations.... In France, the siting of nuclear facilities was not restricted by any regulatory criteria on, say, distance. The safety philosophy adopted in the United States and Great Britain, however, led to strict rules establishing exclusion zones and evacuation zones whose ranges varied depending on the reactor capacity and the surrounding population density. In France, decisions were made on a case-by-case basis and 'the best possible use of distance as a factor of safety' was made."[76]

The policy principles put in place in the 1960s would be revised during the following decade, which was further marked by a reshuffling of the roles played by EDF, the CEA and the public authorities.

2. THE DEVELOPMENT OF FRANCE'S NUCLEAR POWER PROGRAM

Large-scale production of nuclear energy began in the early 1970s, prompting a shift in the landscape of France's nuclear sector. The graphite-gas technology that had been in use since the early days was officially discontinued in 1969. In early 1973, EDF embarked on a vast program of building light water reactors based in design on those already in operation in the United States. A few weeks later, the Central Department for the Safety of Nuclear Facilities (SCSIN) was created within the French Ministry of Industry. Headed by engineers from the Ecole des Mines, the SCSIN was officially in charge of monitoring nuclear facilities. This redistribution of roles was also based on technical dialogue, which was notably used during safety report reviews. The demonstration of safety was based on the concept of defense in depth, which consists in using multiple barriers to prevent accidents or mitigate their consequences. The deterministic method of barriers, which consists in protecting oneself from all plausible adverse events, faced competition from probabilistic approaches derived from recent theoretical work in the fields of industrial reliability and decision-making. The 1975 Reactor Safety Study (WASH-1400) published by the team led by Professor Norman Rasmussen was the first probabilistic safety assessment of an industrial facility. Nevertheless, the use of this tool remained marginal in France,

---

[75] For a former French safety expert, "French cooking is defined as the fact that since all the directors of France's nuclear agencies are graduates of the same school, they are always able to see eye to eye." (interview of December 1, 2007)

[76] Foasso, 2003, pp. 241–242.

where "resorting to 'professional judgment', that is, the expertise internal methods of regulation of the relevant technical experts and professions" remained the prevailing method of handling technical risks[77]. Technical dialogue was at odds with the rationality conveyed by the probabilistic approach.

2.1. Discontinuation of Graphite-Gas Reactors and Start of the Nuclear Power Program

Much has been written about what has been called the "reactor war" (guerre des filières), which pitted the CEA, proponent of gas-graphite reactors, against EDF, advocate of pressurized water reactors, from 1966 to 1969. Although the ins and outs of this war are beyond the scope of this document[78], in December 1969 the French government decided in favor of EDF and authorized the construction of a pressurized water reactor in Fessenheim. The early 1970s marked a major shakeup in France's nuclear power sector. While the entire range of expertise was in the hands of the CEA[79], a foreign technology was brought into the country, with EDF and Framatome, its vessel manufacturer, benefiting from the transfer of knowledge. Accountable for ensuring the safety of its facilities, EDF created two nuclear safety divisions in 1974: one within its thermal production department and the other within its thermal and nuclear studies and facilities department[80].

2.2. A New Direction in Institutionalization: the Expert and the Decision-Maker

While the number of nuclear power plants increased, a division was made between the expert and the policy-maker, with the latter reporting to a new administrative department. This new department drew on the expertise CEA, which retained its nuclear safety assessment functions. In 1976, an institute was created within the CEA with the purposes of providing technical assistance to the French public authorities. Before the start of construction of a facility and its

---

[77] Moatti, J.-P. (1989). *Economie de la sécurité: de l'évaluation à la prévention des risques technologiques*. Paris, La documentation française, p. 7.

[78] This war is discussed in particular by the following two sources: Goldschmidt, B. (1980). *Le complexe atomique. Histoire politique de l'énergie nucléaire*. Paris, Fayard, Hecht, G. (2004). *Le rayonnement de la France. Energie nucléaire et identité nationale après la seconde guerre mondiale*. Paris, La Découverte.

[79] A few years later, the CEA would also be deprived of most of its production activities (mining, uranium enrichment and fuel processing) in favor of the General Company for Nuclear Materials (Cogema), which was created in 1976.

[80] Vallet, 1984.

placement into operation, the experts would be called on to give their opinion before the members of a commission known as an advisory committee.

Emergence of an administrative control authority

The development of France's nuclear power program, spurred by the energy crisis of the 1970s, was cited by the head of the SCSIN as the main reason for the creation of this new administrative agency in March 1973, adding that "a better definition of the respective roles of the CEA and the ministry to which it reports" was necessary. "As we all know," he continued, "the CEA is steadfastly committed to improving nuclear power generation techniques and to developing novel techniques that may be able to compete with those already used in industry. As a result, the CEA cannot be the authority of control that simultaneously determines the safety of techniques, authorizes the use of one, and rejects or restricts the use of another. As has already been said, it cannot both 'judge and party'."[81] A number of sources maintain that industrial companies favored the creation of a government safety agency in charge of preparing of public authorizations. "Allegedly, [they] actually requested to no longer be bound hand and foot to the CEA regarding safety, security, oversight and other matters."[82]

Upon its creation the SCSIN was put under the charge of the engineers of the Corps des Mines of the Ministry of Industry rather than that of the physicians of the Ministry of Health, then in charge of radiological protection,[83] and the engineers of the Corps des Ponts although they participated in the construction of hydroelectric and thermal power stations. The new department had few resources and was staffed by just five engineers. Moreover, although they had long been specialists in the inspection of pressure equipment, the engineers from the Corps des Mines were not specialized in nuclear technology. It was therefore decided that they would draw on the skills of the CEA's specialists when making administrative decisions regarding the granting of authorizations and when inspecting and monitoring facilities. Foasso also notes that it was "the only possible solution given the lack of skilled experts in France's universities, a

---

[81] Taken from an article written by Jean Servant, "La sûreté nucléaire au ministère du développement industriel et scientifique", *Revue française de l'énergie*, n°254, juin 1973, cited by Foasso 2003.

[82] Vallet, 1984, pp. 54–55.

[83] Research and control activities related to radiological protection were under the authority of France's Ministry of Health until 2002.

situation found nowhere else than in France and that set the country apart from other nations such as the United States and the United Kingdom."[84]

Safety experts at the CEA

The CEA thus retained its technical authority over safety. In 1970, Jean Bourgeois became head of the CEA's newly created nuclear safety department. According to Foasso, safety became a full-fledged field of nuclear research in the 1970s. "This is attested by many facts: the number of specialized symposia, the internationalization of the field, which led to the standardization of methods, and the building of a consensus on the rules and practices to be codified."[85] The new department organized itself around this development. As one department manager explained in detail, engineers were either generalist or specialist. Each type had its own role and its own area of expertise: "The generalist engineers are pooled by reactor type (ordinary water reactors, fast neutron reactors and the such) and are in charge of assessing the safety of each facility subjected to review.... These generalist engineers apply the experience they acquire primarily to formulate opinions. They are responsible for constantly updating the acquired body of knowledge on safety for the type of reactor facilities being reviewed, whether it is safety under normal operating conditions or safety in incident and accident situations. In order to do this, the generalist engineers have to know how to maintain close relationships with licensees. Nevertheless, because the technical challenges raised by the facilities require such a large number of disciplines, it is necessary to also have a group of people specialized in these disciplines. These specialist engineers, each of whom possesses the body of knowledge in his or her field, are able to answer the generalist engineers' questions or indicate a deficiency."[86]

This distinction remained in place after the transformation, in 1976, of the department into the Institute for Nuclear Safety and Protection (IPSN). This transformation marked an additional step towards making the safety analysts independent from the CEA. The creation of the SCSIN and the IPSN established clear distinctions between the roles of decision-maker and expert. From then on,

---

[84] p. 261.

[85] pp. 344–345.

[86] Lelièvre, J. (1974). "L'analyse de sûreté et les études correspondantes." *Annales des Mines* (Janvier): 55–61. According to a former safety expert, this organization was also designed "to obtain cross-functional analyses, the lack of which was one of the main shortcomings in the organizational methods in the United States." (interview of December 1, 2007)

the public authorities based their decisions on the opinion of the IPSN's experts while the SCSIN consulted the opinion of experts sitting on advisory committees.

Advisory committees

In order to avoid being subject to the sole discretion of the CEA's analysts, EDF succeeded, as early as 1967, in getting the safety reports for its facilities reviewed by an ad hoc expert committee instead of by the reactor safety subcommission. Chaired by Jean Bourgeois, this ad hoc committee consisted of experts from the CEA, EDF and representatives of the French Ministry of Industry. It was replaced in 1973 by advisory committees. The experts in the CEA's safety department (and the IPSN's experts starting in 1976) would review every application submitted to the Ministry of Industry. These experts would present their findings during advisory committee meetings. The committee's members would base their recommendations, which would have to allow the public authorities to reach a decision, on these findings. The advisory committee's experts would come together at "every step of the safety report – the preliminary version, the provisional version and the final version – presented by EDF for the building and start-up of a reactor as well as whenever modifications were made to a reactor already in operation."[87] The reports would be submitted at various stages: the preliminary report with the application to build the facility, the provisional report with the fueling and reactor ramp-up application and the final report, which gave the results of the operating tests and rules, at the time of commissioning. These three reports were made mandatory by a ministerial direction (27 March 1973)[88], all three reports had to demonstrate the safety of the facility, particularly by proving proper implementation of defense in depth.

2.3. Safety Policies

A number of major international conferences held in the 1950s and 1960s were the scenes of debate on nuclear safety. At the time, the leading roles were held by the Americans and the British, who particularly used the concept of barriers, which are central to defense in depth, a concept used in France's nuclear power plants in the 1970s. Although no one disavowed defense in depth, knowing which incident and accident-initiating events should be included and which should be excluded was problematic. The 1975 Reactor Safety Study (WASH-1400) provided probabilistic answers to these tricky questions. In France,

---

[87] Vallet, 1984, pp. 58–59.

[88] Millet, 1991, pp. 143–144.

however, preference was given to the deterministic approach even if reliability-based reasonings were introduced at this time. The 1970s were a decade dominated by the design and construction of nuclear power plants. Many authors maintain that during this period, engineers, designers and safety experts failed to adequately take into consideration the conditions under which materials were used by licensees as well as how shifts were organized.

Defense in depth

The concept of defense in depth "is not a facility review guide associated with a specific technical solution like a layering of specific barriers but rather a method of reasoning and a general framework for reviewing entire facilities more comprehensively both in terms of their design and analysis."[89] Originally a military strategy, defense in depth consists in using various levels of equipment and procedures in order to maintain the effectiveness of physical barriers placed between radioactive materials and the environment. The principle's simplicity is most likely the reason behind its success. Defense in depth, which commonly consists of several levels of protection[90], was implemented in pressurized water reactors in the 1970s according to the following three levels:

- Prevention of abnormal operation and system failures. *In order to prevent incidents, a high level of quality and reliability must be achieved through the application of proven technologies and standards during design and construction and by providing adequate margins of safety;*

- Control of operation. *Safety or protection systems must be designed to prevent the facility from operating outside its design limits. The most important of these is the emergency shutdown;*

- Control of accidents within design basis. *Safety systems are designed to control postulated accidents.*

Levels 4 (prevention of accident progression and mitigation of the consequences of severe accidents) and 5 (mitigation of radiological consequences of significant release of radioactive materials) were subsequently

---

[89] Libmann, J. (1996). *Eléments de sûreté nucléaire.* Les Ulis, Les éditions de physique, p. 40.

[90] Garrick, B. J. (1992). Risk management in the nuclear power industry. *Engineering safety.* D. I. Blockley. London, McGraw-Hill: 313–346, AIEA (1996). Defence in depth in nuclear safety. INSAG-10. Vienne, IAEA: 33, Libmann, J. (1996). *Eléments de sûreté nucléaire.* Les Ulis, Les éditions de physique, Keller, W. and M. Modarres (2005). "A historical overview of probabilistic risk assessment development and its use in the nuclear power industry: a tribute to the late Professor Norman Carl Rasmussen." *Reliability engineering and System safety*(89): 271–285, Garbolino, E. (2008). *La défense en profondeur: contribution de la sûreté nucléaire à la sécurité industrielle.* Paris, Tec & Doc Lavoisier.

added. The safety functions associated with each of the five successive levels of defense should be "as independent as possible".

The use of three successive barriers to confine radioactive materials inside a reactor core is a well-known illustration of defense in depth. The first barrier consists of metal cladding around the nuclear fuel. The second barrier is a steel vessel housing the reactor core. The third barrier is a reactor containment surrounding the reactor vessel.

Defense in depth is often associated with a deterministic approach to safety. In other words, if the planners consider an accident plausible, it is taken into account in the design of a facility's safety systems. Otherwise, it is excluded and the facility is not designed to withstand this event. F.R. Farmer, who had been a staunch critic of the concept of MCA, is behind the use of the probabilistic assessments in nuclear safety. The following statement by Farmer illustrates particularly well the reason for his criticism: "There is no logical way of differentiating between 'credible' and 'incredible' accidents. The 'incredible' is often made up of a combination of very ordinary events – for example, the breakdown or deterioration that occurs in normal plants and their measuring instruments; and the 'credible' may actually be exceedingly improbable. The logical way of dealing with this situation is to seek to assess the whole spectrum of risks in a quantity-related manner...."[91] The first probabilistic safety assessment of a nuclear facility was made in 1975.

Probabilistic safety assessments

Not only was the probabilistic safety assessment made by Professor Rasmussen's team an important moment in the history of nuclear safety, it also marked the history of reliability, a discipline on the fringes of operational research and engineering sciences. Villemeur and Elms ascribe the development of fault tree analysis, which is central to probabilistic safety assessments, to H.A. Watson and Bell Laboratories. A fault tree starts from a specific initiating event and follows all the possible scenarios that could be caused by this event. The continuation or end of the chain of events leading to an accident is contingent upon the proper operation of the safety systems used. It is therefore necessary to determine the probabilities of success and failure. This is simple when systems are independent of each other and their rate of reliability is known.

---

[91] Cited by Foasso, 2003, p. 235.

The Reactor Safety Study (WASH-1400) was "the first comprehensive assessment of the risk associated with an industrial facility."[92] One of the study's findings was that the probability of a core meltdown was $6 \cdot 10^{-5}$ per reactor per year. This was supposed to make nuclear energy a relatively safe technology. "The summary of the study's findings that is cited the most often is that the chances of a person dying from a reactor accident are about the same as getting struck by a meteorite."[93] The study had many critics in the United States. The American Physical Society (APS) voiced strong reservations about the probability estimates made by Rasmussen's team. In France, the IPSN's nuclear safety director welcomed the study but, like the American experts, remained cautious about the value of the parameters used in the model. Furthermore, the idea that such an assessment could have an impact on nuclear safety regulation created a situation that pitted utility directors against the public authorities. "The licensees cited these studies as a reason for putting a halt to what they considered to be unwarranted safety requests and, most importantly, would not accept probabilistic regulations unless other requirements were discarded. On the other hand, the regulating authority and its technical experts sought protection against beyond-design-basis accidents. They also considered that the operating experience was still not sufficient to warrant a withdrawal of the safety requirements in place."[94]

In addition to implementing defense in depth following the publication of the Reactor Safety Study (WASH-1400), EDF's engineers began thoroughly reviewing the reliability of safety-related systems in 1975. However, the probabilistic approach would be used in France only as an incomplete safety analysis method and not as a full-fledged design method or an instrument of regulation or negotiation between decision-makers, their experts and licensees. According to an account by an EDF official, recorded in the late 1970s, this was primarily due to "the lack of data in areas of very low probability, the difficulties of logical analysis in the areas of common-mode failures[95] and, especially, failures due to human error."[96] This last point is also cited by two representatives of the administrative authority in an article published in the trade journal *La Revue*

---

[92] Villemeur, A. (1988). *Sûreté de fonctionnement des systèmes industriels*. Paris, Eyrolles, p. 15.

[93] Foasso, 2003, p. 312.

[94] Foasso, 2003, p. 323.

[95] Common-mode failures are failure events whose occurrence is dependent on that of another event. Determining the probability of the occurrence of non-independent events is difficult.

[96] P. Bacher « Réflexions sur la sûreté nucléaire », DI 80-01. Note technique, E.D.F., Direction de l'équipement, SEPTEN, pp. 8–9 cited by Vallet, 1984.

Générale Nucléaire: "the inclusion of human error, because of its variety and unpredictability (in relation to the possible deficiencies of an item of hardware performing a given function) does not fit well within the scope of fault tree analysis."[97]

Absence of consideration of human factors

Although it was not easy to include human-induced operating failures in quantitative risk assessments, the deterministic approach was no better in taking them into account. Benoît Journé, in his thesis on the management of nuclear safety, states that during the 1970s "emphasis was placed on the quality of the initial design, the quality of the hardware and the quality of the procedures drawn up by the licensees (Framatome, in France). The facility design engineers had in mind a perfect and entirely self-regulating technical model. ... The role played by people in such facilities was considerably reduced to merely the enforcement of rules. In actual fact, people did not really pose a problem insofar as the engineers were convinced that the technology was safe provided all the automatic controls were left to do their job. ... More than just people, the entire process of operating a facility took a back seat to design. Facility operation boiled down to monitoring the machine as it ran by itself."[98] This situation was not specific to France. The sociologist Charles Perrow has stressed the fact that nuclear power stations in the United States were designed based solely on engineers' performance criteria without notice of the lessons taught by human factors studies, which attempted to include the conditions under which hardware was used by operators. The human factors engineers at Westinghouse, the main designer of nuclear power plants in the United States, were unable to impose their ideas due to the influence wielded by design engineers. As a result, human interfaces were left out of the design of machines. According to Perrow (1983), this led to the development of technical systems that did not take into account the physical and mental characteristics of the people in charge of their daily operation. For Joseph Rees, U.S. regulations in the 1970s took little notice of organizational and institutional issues. "The vast majority of standards (governmental and nongovernmental) concentrated on hardware-related issues – how nuclear plants should be designed and constructed – while hardly

---

[97] Houzé, C. and J.-M. Oury (1981). "L'importance de la fiabilité humaine pour la sûreté des installations nucléaires. L'expérience française et les enseignements de l'accident de Three Mile Island." *Revue générale nucléaire*(5): 419–423, p. 420.

[98] Journé, B. (1999). Les organisations complexes à risques: gérer la sûreté par les ressources. Etude de situations de conduite de centrales nucléaires. *Sciences de l'homme et de la société. Spécialité Gestion*. Paris, Ecole Polytechnique. **Thèse de doctorat**: 434, pp. 111–112.

any notice was taken of the institutional arrangements and processes required to manage, operate and maintain those plants. ... All those matters were largely ignored by a regulatory system consumed by the enormous task of developing hardware standards in response to nuclear power's great construction boom."[99]

2.4. Conclusion: Establishment of the Safety Triad

At the end of the 1970s, France was second only to the United States as the world's most nuclear-reliant country. This increase in nuclear power was accompanied by a clearer distinction in safety roles. These roles were redistributed between an administrative department serving as an authority, a small number of state-owned licensees (EDF, CEA, Cogema) and the IPSN, the assessment institute within the CEA. Thus was formed the nuclear[100] safety triad. However, as Vallet writes, "the safety institution remained ... established within a broad, tightly intertwined network formed by the main organizations involved in nuclear power." The fact that all of these organizations were under the authority of the French Ministry of Industry was a source of much criticism. Simonnot calls it a nucleocracy[101] while Roqueplo describes it as the "monopolization of expertise by the technology's promoters"[102]. The matter of the capture of the regulator and the expert seems particularly relevant in the institutional context of nuclear power in France[103]. Millet denounces the Ministry of Industry's supremacy: "the impartiality in risk handling is compromised by the Ministry of Industry, an 'empire' so powerful that the other ministerial departments have little room for maneuver in this sector. All the organizations involved in the nuclear field are dependent on the ministry. There is a fear that regulation is subordinated to the promotion of nuclear power."[104] Even though, as Hébert said, "In addition to regulations, licensees must meet a slew of requirements", the little weight carried by French regulations fostered a regulatory process predicated on the upholding of a technical dialogue. This was all the more so since "safety rules really don't have any regulatory force (telex, recommendations, fundamental safety rules) .... There is a level of uncertainty

---

[99] Rees, 1994, pp. 21–22.

[100] The term is Foasso's (2003).

[101] Simonnot, P. (1978). *Les nucléocrates.* Grenoble, Presses universitaires de Grenoble.

[102] Roqueplo, P. (1995b). "Scientific expertise among political powers, administrations and public opinion." *Science and Public Policy* **22**(3): 175–182.

[103] Moreover, a safety expert pointed out that licensees sit on the advisory committees. "Some licensees are there to guard their turf." (interview of January 12, 2007)

[104] Millet, 1991, p. 220.

about the legal status of safety rules."[105] Licensees were not alone in seeing an increase in the number of requirements imposed upon them. Design engineers were making more and more demands on operators. The human and organizational aspects that determine the operation of plants were absent from the prevailing policy on the construction of technical facilities. The significance of these aspects would be revealed by the *Three Mile Island* accident.

3. THE THREE MILE ISLAND ACCIDENT AND ITS CONSEQUENCES

The *Three Mile Island* accident in 1979 was a partial nuclear meltdown. Following the accident, President Jimmy Carter set up a commission of experts to conduct an investigation. The commission's report, known as the Kemeny Report for the commission's chairman, listed the main causes of the accident. The accident was also the subject of many publications and analyses[106]. In France, TMI would prompt the creation of an action plan and the formation of committees of specialists from a new discipline. It marked the emergence of human factors in institutions of technical dialogue.

3.1. Causes of the Accident

All the analyses concurred that what initiated the TMI accident was the failure of a relief valve on the pressurizer to close. A light on the control room panel merely indicated that an order had been given to close the valve. It did not indicate whether the valve had actually closed. The operators therefore thought that the valve had close when in fact the primary circuit was being drained of coolant. They then made a series of errors that led to the partial meltdown. By aiming to establish "the fundamental causes of the accident", the findings of the Kemeny Report diverged from those of the initial investigations of the accident, which unanimously blame "operator error".

---

[105] Taken from a speech by J. Hébert at a symposium on nuclear power stations and the environment, cited by Millet, 1991, p. 229.

[106] Devillers, C. (1979). L'accident de Three Mile Island. *11e congrès national de l'association pour les techniques et les sciences de radioprotection.* Nantes, Kemeny, J. (1979). Report of the President's Commission on the accident of Three Mile Island, www.pddoc.com/tmi2/kemeny, Houzé, C. and J.-M. Oury (1981). "L'importance de la fiabilité humaine pour la sûreté des installations nucléaires. L'expérience française et les enseignements de l'accident de Three Mile Island." *Revue générale nucléaire*(5): 419–423, Lagadec, P. (1981). *Le risque technologique majeur. Politique, risque et processus de développement.* Paris, Pergamon Press, Perrow, C. (1984 (1999)). *Normal accidents. Living with high-risk technologies.* Princeton, Princeton University press, Llory, M. (1999). *L'accident de la centrale nucléaire de Three Mile Island.* Paris, L'Harmattan, Walker, J. S. (2004). *Three Mile Island: a nuclear crisis in historical perspective.* Berkeley, The University of California Press.

The two main causes identified by the Kemeny Commission were the lack of operator training and the inadequacy of the emergency response procedures[107]. "... we found that the specific operating procedures, which were applicable to this accident, are ... very confused ...." Furthermore, heavy criticism was leveled at TMI's control room with its maze of lights that lit up and flashed like a Christmas tree or a sign in Las Vegas. The alarms were not classified into categories that would have made it possible to distinguish initiating events from their normal consequences. The relevant information was spread across consoles measuring 15 to 20 meters in length. The computer, overloaded by the information it was receiving, jammed and stopped relaying information for two hours.

Furthermore, "soon after the TMI accident it was discovered that very similar accidents had almost occurred twice before, in 1974 at a Westinghouse reactor in Switzerland and in 1977 at Toledo Edison's Davis Bessa plant in Ohio. Yet there were no meltdowns. In a matter of minutes operators at both plants had successfully diagnosed and solved the problem, thus avoiding serious damage. The TMI investigations also uncovered another telling fact. The power-operated relief valves that failed to close on the TMI reactor, the major culprit in the accident, had failed before – nine times – in reactors of similar design."[108] Blame was thus placed on the lack of training and capitalization on feedback from experiences at other reactors. On this point, Kemeny (1979) and Rees (1994) blame the NRC: "... although the NRC required the nuclear utilities to report all abnormal events, there was no system for distributing these reports to other utilities, no system for evaluating their safety significance, and no system for analyzing their applicability to other nuclear plants."[109] The situation in America made the existence of such a system even more necessary: "... the nuclear utilities led an isolated existence in many significant respects, and that one noteworthy consequence of their isolationism was that hardly any serious attention was given to the operating experience ... of other nuclear plants."[110]

For the Kemeny Commission's experts, the accident revealed "a lack of attention to the human factor". Adjustments were made in both the United

---

[107] Libmann (1996) points out that the procedures did not include leaks of coolant from the top of the pressurizer: "The control room team could therefore rely on neither their training nor a document providing a methodology for identifying the situation. They were alone and in uncharted territory." (p. 191)

[108] Rees, J. V. (1994). *Hostages of each other: the transformation of nuclear safety since Three Mile Island.* Chicago, The University of Chicago Press, p. 22.

[109] Ibid., p. 23.

[110] Ibid., p. 23.

States and France. The concept of defense-in-depth concept was not endangered. As historian J. Samuel Walker writes, "The concept of defense-in-depth, the basic philosophy that guided the regulatory decisions of the AEC and the NRC, was tested as never before. In the face of a massive core meltdown, it worked."[111] Problems encountered during reactor operation were now the object of attention.

3.2. The Accident's Consequences in France

"Given the extent of the French nuclear program, both EDF and the SCSIN were faced with the enormous task of quickly drawing all the lessons of the TMI accident. With the exception of minor points that were easily solved, this review did not reveal any design flaws likely to halt the program or delay the commissioning of nuclear units ready to achieve criticality. However, it did highlight the pressing necessity of paying close attention to operating problems and, more specifically, improving all elements that must allow operators to correctly respond to accident situations."[112] This was the conclusion of the report discussed by the experts of the reactor advisory committee during a four-day meeting in the summer of 1979. An action plan was developed and committees of human factors experts were formed to implement it.

Naturally, this plan was based on the findings of the Kemeny Report. It stipulated the necessary hardware modifications and addressed operator training, the clarity of instructions and information in control rooms and the implementation of operating experience feedback systems[113].

Training of operators in running PWRs began in 1974. Until TMI, operators had merely received general training on how PWRs functioned. The construction of simulators was coming to an end[114]. From then on, "operator training [would include] sessions on a reactor simulator, which [would give] operators a better understanding of how units react during incident situations. The plan is to use

---

[111] Walker, J. S. (2004). *Three Mile Island: a nuclear crisis in historical perspective*. Berkeley, The University of California Press, p. 240.

[112] Fourest, B., Y. Boaretto, et al. (1980). Impact de l'accident de Three Mile Island sur le programme nucléaire français et sur l'analyse de sûreté. *Conférence A.N.S./E.N.S. sur la sûreté des réacteurs thermiques*. Knoxville (U.S.A.).

[113] The analyses of the TMI accident also prompted a significant organizational change with the creation of the safety and radiological protection engineer. We will discuss the issues raised by this reorganization in the following chapter.

[114] Libmann (1996) specifies that "the standardization of the French nuclear power plant fleet made it possible to develop simulators that were directly representative of the varies types of facilities" (p. 194).

the simulator to artificially recreate a certain number of typical accident cases."[115]

Regarding the improvement of instructions and control rooms "considerable effort is being made ... to make operating instructions simpler and clearer. It is important that operators be clearly informed about the state of the unit requiring their attention so that they may be able to save what's essential by avoiding actions that could worsen the situation."[116] The instructions must be reviewed, rewritten and tested on the simulator.

In the control room, the objective was to "make sure that the information made available to operators is reliable, adequately presented and prioritized. In particular, there are plans to improve the presentation of alarms, core temperatures and information on the state of safety-related valves."[117] A better presentation of information was sought, some measurement ranges were broadened, new indications were added and the alarms were prioritized. Essential information was displayed on a safety panel.

It was also necessary to "analyze certain moderate-frequency incidents that could lead to more severe accidents. Such incidents are factored into the design process and encountered during operations. Testing and operating experience can reveal weaknesses in facilities. The situations caused by these weaknesses should be analyzed. An operational experience feedback system ensures that information on test and operating results is reported to the designer. Moreover, EDF has been requested to perform and communicate to the safety authority a biannual assessment of the main incidents that occur, the analysis of such incidents, the lessons learned and the arrangements made for each unit."[118]

3.3. Conclusion: Emergence of a New Field of Specialization

The licensee and the IPSN called on human factors experts in particular to implement these actions and assess their impact on safety.

A human factors laboratory was created at the IPSN. The laboratory's experts had a dual function: participate as assessors in safety analyses conducted by the IPSN and serve as advisers to designers and licensees. Regarding this matter, allow us to point out that Maurice Gomolinski, the physicist appointed head of

---

[115] Fourest, B., Y. Boaretto, et al. (1980). Impact de l'accident de Three Mile Island sur le programme nucléaire français et sur l'analyse de sûreté. *Conférence A.N.S./E.N.S. sur la sûreté des réacteurs thermiques.* Knoxville (U.S.A.).

[116] Ibid.

[117] Ibid.

[118] Ibid.

the laboratory, initially hoped to collaborate with the licensee through a joint partnership between EDF and the CEA. EDF's refusal led to the creation of two groups: one at EDF in 1982 and another at the IPSN's human factors analysis laboratory one year later. However, this situation would not prevent the teams at the CEA and EDF from conducting "many joint studies."[119] The job of the dozen experts in the IPSN's new laboratory varied between that of assessor and adviser. In a way, this dual-hatting was the result of a historical process characterized by "an organizational internalization of risk management", the institutionalization of technical dialogue that placed greater emphasis on the exploration of knowledge than on the standardization of practices and oversight based on procedures. It might also have been necessary for addressing human factors issues.

---

[119] Gomolinski, M. (1986a). La prise en compte du facteur humain dans la conception et le fonctionnement des centrales à eau légère, CEA/IPSN/DAS/LEFH, p. 5.

Chapter 2. Incorporation of Human Factors in Assessment Processes

How did the experts in the human factors analysis laboratory, administered by the safety analysis department, "mix in" with the institute's experts who conducted safety analyses falling under so-called technical fields? What issues did they have to address? How was their activity organized within the technical dialogue institutions?

A historical analysis[120] shows a significant change in their position within the institute as well as in their technical relationships with licensees. Today, the main task of human factors experts is to perform assessments – first referred to as analyses then evaluations. The institutions where these activities are performed have changed. The Chernobyl accident was the starting point of a "long road to independence and transparency" and culminated in the separation of the assessment institute from the CEA and the creation of an independent administrative authority in charge of nuclear safety. It is against this institutional background that the human factors experts conducted their analyses by collaborating not just with other experts at the institute but also representatives of the French nuclear safety authority, the ASN.

---

[120] This historical analysis is based on the archives of the institute's human factors experts and interviews. We compiled and classified the various issues addressed since the entity's creation and conducted some thirty interviews of important figures in its history. The nuclear safety institutions were updated after the reading of a number of reports and publications. The main organizational arrangements of the human factors assessments discussed on the following pages were identified while observing, in real-time, a number of analyses conducted by the institute's experts.

1. GENEALOGY OF THE WORK OF THE INSTITUTE'S HUMAN FACTORS EXPERTS

The retrospective analysis of the work of the institute's human factors experts shows a fundamental change. In the wake of TMI, the experts worked together in a laboratory, an apt term since the work of its experts consisted primarily in carrying out studies. Starting in the late 1980s, under the impetus of their new head, the experts began focusing primarily on safety evaluations. This important shift, exemplified in particular by the conversion of the laboratory into a department, was more in line with the main activities of the new department's parent department. It was from this moment that the human factors experts began conducting three types of evaluations that continue to make up the core of the department's activities.

1.1. The Laboratory Years

In the 1970s, a few engineers turned their attention to analyzing human errors that could occur during the nuclear power plant operation. The TMI accident was a catalyst and its events were the first that the human factors experts analyzed. They were also able to turn their attention to relatively varied range of issues by conducting studies in nuclear power plants as well as laboratories and fuel cycle plants. Lastly, they took part in a number of safety analyses that culminated in the formulation of recommendations that were discussed during advisory committee meetings. Their participation in these analyses made it difficult for them to gain access to facilities. And yet this access was crucial in order for them to be able to collect the data required for the analyses and the studies. This difficulty contributed to frustrating the efforts of the institute's human factors experts.

In the late 1970s, the IPSN's engineers were using analysis tools derived from research in reliability and decision-making to study incidents. During a conference in 1984, Gomolinski, then in charge of the human factors analysis laboratory, discussed the method used by the experts, describing it as a "rather classic method called 'event tree analysis' in which you start from the descriptions of an actual incident and, using a slew of ramifications, work your way down to its causes. It has nothing specific to do with human factors, but let's say that in the case at hand we specifically looked at the human factor branches and tried to provide explanations."

In 1982, an engineer at the institute who had a keen interest in incident analyses, was hired away by EDF to participate in the creation of groups of human factors experts. Despite this transfer of skills, incident analysis would remain a significant part of the work of the institute's experts. "We constantly track incidents involving human errors," said its manager in 1986. "How they are tracked depends on the magnitude of the incident. At the very least they are entered in a special database, while some require on-site analyses."[121] In a programmatic memo written in 1982, Gomolinski emphasizes that decisions to conduct on-site incident analyses were made with EDF's human factors experts.

This interest in incidents made it possible to set into motion a certain number of 'corrective actions' and become aware of potential problems. "A special group at EDF ... reviews the lessons of the most significant incidents .... These feedback reviews lead to many general studies being conducted on a plant designated to be used as a pilot. The [IPSN] is often associated with these on-site studies."[122] In the same memo, Gomolinski explains in detail the main studies conducted with EDF, i.e. a study of local identification errors (mistaking of devices within a unit or between two twinned units), a communication study (study of communication tools and vocabulary), a stress study and a study of lineup errors, (errors during circuit reconfiguration). For certain topics, the experts collaborated with external researchers, particularly those at the industrial and organizational psychology laboratory of the Ecole pratique des hautes études (EPHE).

During the same period, a number of studies were being conducted on the Osiris experimental reactor operated by the CEA and a Cogema plant. During a conference in January 1984, Gomolinski explained how one of these studies was conducted: "Osiris is a fission research reactor operated by the CEA to irradiate items intended for use in hospitals and various industries. Because it operates 24 hours a day, its on-site [human factors experts] first must look at how shifts are organized within it. This allows them to identify a number of issues that would be worth addressing. Obviously, not all aspects are of equal interest when studying normal operation. However, the experts identify a number of areas that merit closer attention. The workstations to be studied are chosen with the consent of the main parties concerned.... Now, there is one thing that is really interesting to see in a control room. It contains a certain amount of equipment for displaying pressure levels, temperature readings and other such information. This equipment was installed according to the designer's specifications. What is interesting is to compare the designer's ideas, which are implicitly present in the arrangement of the equipment in the control room, with how important the operators consider them to be. This is done by observing the operators during time intervals to see what equipment they watch both during normal operation

---

[121] Gomolinski, M. (1986b). Paramètres humains dans la sûreté des installations nucléaires, CEA/IPSN: 10, p. 2.

[122] Gomolinski, M. (1986a). La prise en compte du facteur humain dans la conception et le fonctionnement des centrales à eau légère, CEA/IPSN/DAS/LEFH, p. 9.

and operations such as reactor startup. And what you get are surprises. You notice that the equipment that is watched the most is not necessarily the equipment you would have thought. Another method consists in taking operators outside the control room and asking them to redraw the plant overview panel – there usually is one in a control room – exactly how they would design it. Very curious things happen. I mean, what they draw is usually different from what they work with and very often they come up with similar ideas. You feel the need to have such or such information that does not appear on the plant overview panel". Conducting a study thus requires close collaboration with reactor staff and consultation with facility managers[123].

During this period they also contributed to a few analyses within the safety analysis department, which administered the laboratory. The training of operating and maintenance personnel was thus analyzed. However, control room ergonomics received the bulk of the attention of the human factors experts. The experts found "many flaws in the presentation of information on the panels (arrangement, legibility), the arrangement of control devices on the consoles, the way in which procedures were written, the type of information required to properly follow these procedures, etc ...." in the control rooms of the 900 MW plants in operation at the time.[124] "This work was performed after interviewing operators and building a mock-up."[125] The findings of these analyses were used to design the control rooms of other series of reactors (particularly those of the 1300 MW reactors).

Furthermore, in 1980, EDF made the decision to computerize the human-machine interfaces in the control room of its N4 reactor series (1450 MW). In the 1980s the institute's human factors experts participated in this project alongside as members of a multidisciplinary group that included occupational physicians, ergonomists, control experts and statisticians. One of the group's purposes was to analyze tests by operators on a full-scale simulator. Yet again, "specific findings were obtained, particularly regarding workstation ergonomics, the use of the various types of images made available to operators, how well they were outlined, codings, the handling of alarms and the design of their presentation and management system, aspects related to information searching

---

[123] According to Gomolinski, some studies were conducted "at the request of facility managers."

[124] Gomolinski, 1986a, p. 2.

[125] Ibid., p. 6.

and collection, the organization of work within the team."[126] In a retrospective study on the participation of the institute's human factors experts in this work, which lasted for several years, Tasset points out that, due to the experts' (over)involvement in the project, at the time it was "preferable to present [their report] ... before the advisory committee simply as one of informational purposes rather than as a safety analysis in the strict sense."[127] This decision, viewed by some as muddled, was a main factor that led to the crisis that hit the laboratory in 1988.

Being able to conduct a study properly required obtaining the licensee's consent. Thus, in order to conduct an incident analysis, the department head had to negotiate with his or her counterpart at EDF. Likewise, in order to analyze the normal operation of a laboratory or plant, he or she had to attract the facility manager's interest. Some licensees sought such cooperation as it could be a source of good advice. However, others saw it differently when the advice given was liable to turn into a requirement. But, as Gomolinski points out, "although these studies are not directly related to the performance of a safety analysis, their general nature often makes them a valuable technical resource"[128]. Thus, licensees viewed these studies as threats despite their explicit objective: to build up a body of knowledge to improve safety. Basically, for them it was "today a study conducted by experts, tomorrow a request by the authority", with the possibility of such request running counter to their interests. The laboratory's experts were therefore invited to less and less studies.

The laboratory for the study of human factors continued to primarily publish reports, which, in the words of Gomolinski's successor during an interview, "sat there gathering dust." And when the institute's human factors experts were able to participate in safety analyses, such as for the project to design the N4 series' computerized control room, they seemed to have a hard time taking a stance. Their direct participation in the project, which perhaps was particularly conducive to the acquisition of knowledge, turned out to be incompatible with an objective of evaluating and analyzing safety. These difficulties in gaining access to facilities and taking an adequate stance in order to analyze safety were compounded by the loss of experts lured away by enticing career opportunities from licensees and drove the laboratory into a crisis. Of the eight

---

[126] Oudiz, A., E. Guyard, et al. (1990). Gestion de la fiabilité humaine dans l'industrie nucléaire, quelques éléments. *Les facteurs humains de la fiabilité dans les systèmes complexes.* J. Leplat and G. d. Terssac. Marseille, Octarès: 273–292, p. 284.

[127] Tasset, D. (1998). "Palier N4: évaluation de la sûreté des aspects facteurs humains de la salle de commande informatisée." *Revue générale nucléaire*(1): 20–26, p. 21.

[128] Gomolinski, 1986b.

experts at the laboratory in 1985, only two remained by Gomolinski's side in 1988. And Gomolinski was already on his way out.

1.2. From Laboratory to Department

Maurice Gomolinski had decided to allocate the bulk of the laboratory's resources to the performance of studies. For André Oudiz, the engineer who succeeded Gomolinski in 1988, this decision was a mistake. Oudiz shifted the focus of the laboratory's operations to safety analyses and cut back on studies. "In order to survive, we had to make our work useful for our colleagues who were conducting safety analyses. It was necessary that we, too, conducted analyses." To solve the problems of gaining access to facilities, Oudiz did not hesitate to stand up to licensees. "If we were going to provide relevant recommendations, we needed access to facilities. I told EDF's people that if they didn't let us access their facilities, we would be harder with them and provide less relevant recommendations."

Oudiz's strategic showdown worked. Human factors henceforth became a specialized area like the rest in the safety analysis department. "We went from a situation where we did 80% studies and 20% assessments to an inversely proportional one of around 75% assessments and 25% studies."[129] This increase in the annual number of safety evaluations was accompanied by the emergence of a new type of analysis devoted to more organizational topics. This can be understood as influenced by analyses and commentaries on accidents in the nuclear, aerospace, chemical and oil industries[130] that stressed the organizational, institutional and cultural shortcomings that led to these events. The success of the safety culture concept[131], defined in the aftermath of the 1986 Chernobyl disaster, illustrates this growing concern for determining factors of safety that would no longer be purely technical or purely cognitive.

---

[129] Oudiz, A. and G. Doniol-Shaw (2005). Histoire de l'ergonomie dans les entreprises: l'IPSN. *Bulletin de la société d'ergonomie de langue française*. Paris.

[130] Shrivastava, P. (1987). *Bhopal: anatomy of a crisis*. Cambridge, Ballinger, Feynman, R. P. (1988). "An outsider's inside view of the Challenger inquiry." *Physics today*: 26–37, Lagadec, P. (1988). *Etats d'urgence. Défaillances technologiques et déstabilisation sociale*. Paris, Seuil, Paté-Cornell, E. (1990). "Organizational aspects of engineering system safety: the case of offshore platforms." *Science* **250**: 1210–1217, Reason, J. (1990). "The age of organizational accident." *Nuclear engineering international*: 18–19, Vaughan, D. (1990). "Autonomy, interdependence, and social control: NASA and the space shuttle Challenger." *Administrative science quaterly* **35**: 225–257.

[131] For an in-depth look at this notion in the area of industrial safety, cf. Chevreau, F.-R. (2008). Maîtrise des risques industriels et culture de sécurité: le cas de la chimie pharmaceutique. *Sciences et génie des activités à risques*. Paris, Mines ParisTech. **Thèse de doctorat**: 285.

One aspect specific to France is that all of the country's electricity is produced by EDF, which opted to standardize its nuclear units. Because certain policies and managerial decisions implemented by EDF can affect the utility's entire fleet of plants, topics specifically related to human and organizational factors are evaluated at all of EDF's nuclear power plants. As a result, several analyses of the organization of the control room teams at EDF's reactors were conducted between 1991 and 1997. Without going into detail about the experts' work, we will describe on the following pages the decisions that led to the start of the evaluation.

Until then, and in a nutshell, control room teams worked in rotating shifts. Each shift was led by a supervisor who served as both shift manager and real-time production manager. At the same time, there was also a safety and radiological protection engineer[132] who was outside the control chain of command and whose sole task was to provide an additional level of facility safety monitoring. A key change introduced was to place the shift supervisor in charge of safety within the facility. Because it considered that the presence during shifts of both the shift supervisor and the safety and radiological protection engineer would lead to confusion as to who was in charge of safety, EDF removed the engineers from its shifts and reorganized the daytime shifts at its reactors.[133]

Between 1990 and 1996, three analyses were devoted to maintenance operations. One of them addressed in particular "the impact of the policy of entrusting safety at EDF facilities to contractors". This determination to assess how organizational, managerial and policy decisions impact safety has been illustrated in recent years by a number analyses, most notably in 2003 (organization of radiological protection), 2005 (management of skills and qualifications of operating personnel) and in 2007 (safety management).

Like experts in other areas of nuclear safety, human factors experts have been contributing to safety evaluations of facilities operated by the CEA and Areva since 1990. Between 1990 and 2004 they participated in twenty-nine safety reviews. The topics reviewed fall within the entire spectrum of human and organizational factors. Pierre Boutin, who succeeded André Oudiz, listed them in a presentation before the SEFH: "instructions, workstation ergonomics, physical and mental aspects, leadership, motivation, group dynamics, human-

---

[132] A position that, as stated earlier, was created following the TMI accident.

[133] Charron, S. and M. Tosello (1994). *Ergonomie et évaluation de sûreté dans le secteur du nucléaire*. XXIXe congrès de la société d'ergonomie de langue française, Eyrolles, Journé, B. (1999). Les organisations complexes à risques: gérer la sûreté par les ressources. Etude de situations de conduite de centrales nucléaires. *Sciences de l'homme et de la société. Spécialité Gestion*. Paris, Ecole Polytechnique. **Thèse de doctorat**: 434.

machine interfaces, duties, responsibilities, cooperation, communication, skills management, training, decision processes, management, safety culture, feedback, methodology, etc."[134]

1.3. Conclusion: the Department's Work

This analysis of the archives of the institute's human factors experts showed that its activities have changed. Human factors, an emerging field in an organization tasked with producing safety evaluations, could hardly remain a field reserved solely for the production of knowledge. Like most experts at the institute, the experts in the SEFH issued opinions and assessed the safety of the facilities operated by licensees. The analysis of the entire range of analyses performed since 1989 highlights the following three types of assessments:

• *incident analyses, which can be conducted at any INB (basic nuclear facility);*

• *participation in safety reviews of research facilities or fuel cycle facilities;*

• *cross-disciplinary assessments in EDF nuclear power plants.*

In order to assess safety at facilities, the experts must take into consideration policy and managerial aspects as well as organizational factors. This increase in the scope of their discipline was notably the result of the strong repercussions of several industrial accidents, including the Chernobyl disaster. This accident, the worst to date, also contributed to prompting profound changes in nuclear safety institutions. Before explaining how human factors assessments are organized, the institutional context in which these assessments are carried out requires updating.

2. RESTRUCTURED NUCLEAR SAFETY INSTITUTIONS

The Chernobyl accident prompted major reforms of institutions that assess and regulate nuclear facilities. The magnitude of the radioactivity released and unanimous criticism of announcements issued by various governments raised public outcry. The values of transparency and independence were demanded by many spokespersons from the worlds of politics and science as well as associations. In the 2000s, these values would be translated into actions with the creation of the IRSN and the adoption, by the French parliament, of the first nuclear law, which officially created what was wrongly referred to as the Nuclear Safety Authority (ASN). France's nuclear safety institutions were thus brought into line with international standards.

---

[134] Boutin, P. (2001). L'expertise des facteurs de performance humaine dans les installations nucléaires, CEA/IPSN/DES/SEFH: 14.

2.1. How the Chernobyl Disaster in France was Managed and How its Management Turned to Disaster

The worldwide consciousness prompted by the disastrous health consequences of the Chernobyl accident was the opportunity for the IAEA to make its voice better heard: "national borders are absolutely no protection against radioactive poisons."[135] It created the international nuclear safety advisory group (INSAG), a group of experts whose reports would spark growing interest on the part of licensees. In France's safety institutions, accident analysis could only intensify the importance of human factors and, as we have already said, would contribute to expanding the scope of this specialized area. Furthermore, notice was paid to the management of emergency situations caused by severe accidents[136]. This last point was justified in particular by the disastrous handling of the accident by France's public authorities and the Ministry of Health's Central Safety Department for the Protection Against Ionizing Radiation.

In the wake of this unprecedented crisis, France's National Assembly took up the matter. "The crux of the debate was the creation of an independent national nuclear authority that would regulate the safety of nuclear facilities. But the government frowned on this solution, so it was unanimously decided that the French Parliamentary Office for the Evaluation of Scientific and Technological Choices would issue a public report on nuclear safety each year."[137] This office, or OPECST, was also charged with analyzing the incident and its handling in France[138]. In particular, the report made a case for increasing the independence of the IPSN's experts from all the licensees. This included the CEA, under which it was still operating at the time.

2.2. "The Long Road to Independence and Transparency"

This desire for transparency and independence would become the guiding theme of reforms recommended a decade later by member of Parliament and chairman of the OPECST Jean-Yves Le Déaut in his 1998 report titled "The Long Road to Independence and Transparency". These reforms would be enacted in

---

[135] Foasso, 2003, p. 441.

[136] Dupraz, B. (1986). "La prise en compte de l'expérience pour maintenir et améliorer la sûreté des centrales nucléaires." *Annales des Mines*: 41–46, Libmann, J. (1996). *Eléments de sûreté nucléaire*. Les Ulis, Les éditions de physique.

[137] Millet, 1991, p. 62.

[138] Raush, J.-M. and R. Pouille (1987). Conséquences de l'accident de la centrale nucléaire de Tchernobyl et sûreté et sécurité des installations nucléaires, Office parlementaire d'évaluation des choix scientifiques et techniques.

the 2000s. Déault's report advocated severing the ties between the assessment institute and the CEA and made the following three recommendations:

- *"Because the issues of safety and radiological protection are closely intertwined, radiological protection must be addressed on the same level as safety, as is the case in other countries such as Great Britain and Germany.*

- *A distinction must be maintained between authority and assessment.*

- *A safety expert cannot be in the service of a licensee or a major research organization that advocates nuclear energy. The IPSN and the CEA must be administratively separate."*[139]

Three years after the report, these recommendations were incorporated in the law of May 9, 2001. The Institute for Radiological Protection and Nuclear Safety (IRSN), a public corporation separate from the CEA, was created by Article 5 of this law. The IRSN was placed in charge of research and assessment activities in the field of nuclear safety and radiological protection. The Office for Protection against Ionizing Radiation (OPRI), which had replaced the Service for Protection against Ionizing Radiation in 1994, was dissolved.

The history of the IRSN is closely tied to that of the ASN, whose gradual and steady rise in influence has marked developments in France's nuclear power industry since the creation of the SCSIN. The SCSIN's conquest for status is summarized in a 2005 report from the Cour des Comptes, France's national auditor: "Created in 1973, the SCSIN became the Directorate for Nuclear Facility Safety (DSIN) in 1991 and was made the General Directorate for Nuclear Safety and Radiation Protection (DGSNR) in 2002. It therefore does not have the status of an administrative authority. Despite this, for over ten years it has been billing itself as the French Nuclear Safety Authority, or ASN. As a result, this central government department has been acting, apparently unhindered, as if it were an independent authority. This attitude has been facilitated by a remarkable continuity in that the same person has been at its helm since 1993. This situation is causing concern among those who fear that the authority's independence is more a reflection of its director general's personality than of its status."[140]

The little central safety department created in 1973 thus kept on getting bigger. "The number of INB [basic nuclear facility] inspectors rose from a little

---

[139] Le_Déaut, J.-Y. (1998). Rapport sur le système français de radioprotection, de contrôle et de sécurité nucléaire: la longue marche vers l'indépendance et la transparence, p. 2.

[140] Cour_des_comptes (2005). Le démantèlement des installations nucléaires et la gestion des déchets radioactifs. Rapport au Président de la République suivi des réponses des administrations et des organismes intéressés. Paris, Cour des comptes: 279, p. 25.

under 30 in 1980 to more than one hundred in 1990 ..., the number of yearly inspections increased from nearly 200 in 1980 to nearly 500 in the early 1990s."[141] The scope of its activity grew when it was put in charge of regulating radiological protection in 2002. The power it seemed capable of wielding over the electrical utility EDF also grew. For example, in an unprecedented media coup, it placed the Dampierre site under close surveillance in 2000. Lastly, its international clout grew when, under the impetus of its chairman André-Claude Lacoste, it created the Western European Nuclear Regulators' Association (WENRA).

In 2006 a watershed moment occurred when the department acquired *de jure* the independence the report by the Cour des Comptes had *de facto* ascribed to it. Like many regulatory bodies, the ASN was an independent administrative authority.

2.3. Conclusion: a New Institutional Model?

The rise in power of a newly independent authority[142] a certain level of neutrality with regard to licensees; the adoption of the first law on nuclear energy; no-holds-barred real-time distribution of reports, minutes and releases. These are as many powerful messages that inform a public continuously searching for the truth and legitimize industrial and research activities in a field affected by historic accidents. These reforms make the organization of nuclear safety an institutional model of risk management. Thus, for Claude Gilbert, "one of the most significant changes – and one of the most innovative as well in risk management – took place in the nuclear sector. This change was the result of the clear distinction that was gradually made in the organizations directly in charge of hazardous activities, the regulatory authorities and the assessment bodies .... Because of the manner in which this change gradually took place, the system is based on placing the main bodies in adversarial roles, whereby each acts and reacts according to its own logic and interests. A certain desire for transparency and media coverage of the disagreements between the regulatory

---

[141] Foasso, 2003, p. 523.

[142] Nevertheless, independence is a value often examined by researchers interested in assessment and risks (Decrop and Galland 1998; Godard 2003; Barthe and Gilbert 2005 are a few examples). Olivier Godard clarifies this idea by denouncing the illusions it creates: "The independent expert – that is, the warrior monk of the Republic who has no attachments to material or personal gain but is able to gain access to knowledge without bias of any kind – does not exist. The aim should not be the independence of experts, but rather the independence of collective assessments. This is all the more true when independence is likened with anti-establishment views." (2003, p. 10)

authority and the main licensee contributed to establish this system, which is the foundation of the bulk of nuclear safety policy in France."[143]

Although "the long road to independence and transparency" has not ended, it has played a role in reshaping the institutional landscape in France. It has done so by erasing certain aspects that were once specific to France (such as the Ministry of Industry overseeing these institutions and the lack of parliamentary debate) and by modeling itself more on international standards. This is the institutional context in which nuclear safety assessments are produced, particularly in the field of human factors.

3. THE ORGANIZATION OF HUMAN FACTORS ASSESSMENTS

Although the human factors experts work in the SEFH, which is part of the Reactor Safety Division (DSR), they also work on matters relating safety in plants and laboratories. The coordination between the human factors experts and the other experts at the institute depends on the type of case. The representatives of the ASN also take part in the process of producing human factors assessments. Human factors assessments are therefore akin to an organized industrial activity in that they involve multiple parties and require structured coordination.

3.1. The Department for the Study of Human Factors (SEFH)

Following the departure of André Oudiz in 1994 and Pierre Boutin in 2003, François Jeffroy was appointed head of the Department for the Study of Human Factors. Before joining the institute in the mid-1990s, Jeffroy had worked in the field of software ergonomics[144]. His first task at the institute was to participate in assessing ergonomics in the control room of the N4 reactor series. He then worked for several years as a generalist engineer in charge of assessing the safety of research laboratories and fuel cycle plants. Before Jeffroy was appointed head of the SEFH, the turnover rate in the department had spiked. Only two members of staff had been in the department for a meaningful length of time. As a result, two new experts were hired.

Ergonomists and psychologists make up the department's body of human factors experts. There is also a sociologist, whose presence in the department is a reflection of the wider range of issues addressed by the department. The staff

---

[143] Gilbert, C. (2002). "Risques nucléaires, crise et expertise: quel rôle pour l'administrateur?" *Revue française d'administration publique* **3**(103): 461–470.

[144] Theureau, J. and F. Jeffroy (1994). *Ergonomie des situations informatisées. La conception centrée sur le cours d'action de l'utilisateur.* Toulouse, Octarès.

in the department can be split into two groups: young men and women with post-graduate degrees and seasoned professionals with backgrounds in other industries, particularly aeronautics. A former shift supervisor at a nuclear power plant assists the experts, particularly in analyzing incidents at EDF reactors. Lastly, the department has worked with contractors. For example, in 2007, it hired two human factors consultants to participate in safety assessments conducted by the department. The staff level in the department has remained stable since 2004. At the time of publication of this book, there were a dozen human factors experts working on safety assessments.

It must be acknowledged that few policy elements are made within the department. It would therefore seem that a large part of job training takes place in the field[145], with experts building on a base of knowledge gleaned from university studies or past work experience. That said, technical meetings, informal discussions and in-house seminars provide opportunities to compare viewpoints, and discuss the process and conclusions of assessments as well as findings published in the literature. Furthermore, the department is organized into several skills hubs to allow each expert to make progress on specific cases. In the case of EDF reactors, some experts examine feedback using incident analyses while others address operational issues and others look at management issues. Two experts focus on safety reviews of research laboratories and fuel cycle facilities and one expert focuses on safety reviews of experimental reactors. Altogether, more than a dozen assessments are conducted every year.

In addition to conducting safety assessments, the head of the SEFH endeavors to develop the "study and research" side, particularly in order to design policy principles and strengthen the knowledge used for assessments. This determination is materialized in particular by the development of thesis subjects with universities. Paradoxically, the subjects studied have nothing to do with nuclear power[146]. Relations between the human factors experts and licensees of basic nuclear facilities are structured by the safety assessment, which have remained the department's main activity since implementing Oudiz's

---

[145] This is one of the points highlighted by Céline Granjou's research on several assessments, including one focused on the risks related to the use of GMOs (cf. Barbier, M. and C. Granjou (2003). Experts are learning. *EGOS*. Copenhaguen, Granjou, C. (2004). La gestion du risque entre technique et politique. Comités d'experts et dispositifs de traçabilité à travers les exemples de la vache folle et des OGM. *Sociologie*. Paris, Université Paris 5 René Descartes. **Thèse de doctorat**: 488.

[146] One thesis recently defended draws on un groundwork conducted in the field of chemistry (Colmellere, C. (2008). Quand les concepteurs anticipent l'organisation pour maîtriser les risques: deux projets de modifications d'installations sur deux sites classés SEVESO 2. *Sociologie*. Paris, Université de technologie de Compiègne. **Thèse de doctorat**: 409). Another thesis currently being written deals with rail transportation.

strategy. The distribution of the department's resources that are currently dedicated to the various assessment types presented in the conclusion to the section one is provided in Table 2.

	FACILITY-SPECIFIC ASSESSMENT		CROSS-FACILITY ASSESSMENT
	Safety review	Incident analysis	
Nuclear power plants		+ +	+ + +
Research facilities and fuel cycle plants	+ + +	+	

Table 2: Distribution of the resources in the SEFH assigned to the performance of the various assessment types (proportional) The organization of the production of human factors assessments.

The organizational conditions defined in the institute for the production of human factors assessments are specific to each of the three types presented.

3.2. Organizational Models of the Production of Human Factors Assessments

As stated earlier, both of the institute's departments in charge of conducting safety assessments of basic nuclear facilities (the DSR and the DSU) are staffed by generalist and specialist engineers. Some of these generalist engineers are assigned to one or more facilities and are known as coordinators. When an assessment is conducted at a facility, it is overseen by its coordinator. This is the case of incident analyses and safety reviews. Some cross-disciplinary assessments, which are conducted primarily at nuclear power stations, are led by the experts of the SEFH.

If an incident occurs, the licensee must notify the safety regulator's representatives. It then has forty-eight hours in which to report the event to the regulator and two months in which to provide it with a report on the incident. Usually, the institute is informed at the same time and the facility's coordinator is often one of the first parties informed. The circumstances and severity of the incident may prompt the institute to conduct an in-depth analysis. When human or organizational factors are blamed, the coordinator may request support from the SEFH, in which case the generalist engineer's supervisor contacts the head of the SEFH to request that it lend one of its experts. If the head consents to do so, a human factors analysis is initiated. The coordinator and the human factors expert discuss the objectives of the analysis and the means to be used. These initial contacts lead to the drafting of a request officializing the request of the generalist engineers. After the analysis is completed and written, the human factors expert submits his report to his department head for proofreading, comments and amendments. Sometimes several drafts are required before the final version is submitted to the coordinator's supervisor for inclusion in his own analysis. Then, after discussions between the generalist and specialist engineers and validation by management, the analysis may be read and commented by the licensee before being sent to the ASN.

It is the managing engineer's job to lead the vast undertaking that is the safety reference framework[147]. He can call on more than a dozen experts, including human factors experts, to help him. The process of validating and transmitting the various contributions are identical to those used during an incident analysis. Once the coordinator receives the contributions, he adds them to a report that must be validated by his management before being distributed to the members of the advisory committee. The coordinator is the rapporteur of the meeting held on the case. The licensee's representatives attend this meeting. The human factors expert may be called on to present his findings and draft recommendations. In the event of a disagreement, the licensee presents his viewpoint. This is followed by deliberations among the members of the advisory committee. They may choose to accept, modify or reject the expert's proposal. After the meeting they draw up an opinion and send it with the recommendations to the ASN.

Assessments that are designed to assess the impact of managerial policies on safety at a group of facilities or the ergonomic design of a new control room are usually led by the SEFH. Several human factors experts participate in the analysis of documents and the collection of data obtained during interviews and in-situ observations. Other generalist or specialist experts at the institute may also be called on.

The organization of human factors assessments thus depends on the type of case at hand. Table 3 presents the values taken on by the three assessment types presented based on the "organizational variables" included in our descriptions.

---

[147] This baseline document usually consists of a safety analysis report, general operating rules and an on-site emergency plan.

	Incident analysis	Contribution to a safety review	Cross-disciplinary assessment
Number of HF experts involved	Just one and the head proofreader	Just one and the head proofreader	Several, including the head proofreader
Role of the SEFH	Contributor	Contributor	Coordinator, occasionally Contributor
Validation of the assessment	By the generalist engineers then by management	By the generalist engineers then by management	By management
Other experts at the institute who are involved	Coordinator (and other experts, if need be)	Coordinator (and other experts, if need be)	Contributing generalist or specialist experts, if need be
Examination by the advisory committee	No	Transmission of the analysis and presentation, if need be	Transmission of the analysis and presentation

Table 3: Organization of the production system based on the type of assessment.

3.3. The ASN's Role in the Production of Human Factors Assessments

The ASN has been implementing a human factors action plan under the responsibility of a human factors expert since 2004. A human factors assessment must be conducted in EDF's facilities once every two years. The ASN's representatives also asks that the institute's experts to include human factors assessments in the safety reviews of the CEA's laboratories and experimental reactors and Areva's plants.

The ASN thus plays a role in the selection of topics and facilities for assessment. Future assessments are decided during meetings attended by representatives of the ASN and the IRSN. Obviously, the resources to be allocated for these assessments must be managed. Once the assessment has been decided, the ASN's representatives contact the licensee to request the documents that will contain the main data to be assessed. Then, with the institute's experts, they determine the eligibility of the case. Any discussions that take place focus on initiating the assessment. The ASN officially refers the case to the advisory committee and the assessment institute. One or more representatives of the ASN monitor the assessment and remain in contact with the managing experts. They may participate in certain meetings held over the course of the assessment (kick-off meeting, halfway meeting, preparatory meeting for the advisory committee meeting). Lastly, the ASN is responsible for formulating and tracking requests resulting from the recommendations of the IRSN's experts and the advisory committee. The implementation of these requests is checked during future assessments or inspections. In the production of human factors assessments, the ASN's representatives have an influence on what cases are selected, participate in the coordination of cases, formulate requests to licensees following assessments and monitor the performance of assessments.

3.4. Conclusion: Collective and Procedural Assessments?

This description of the organization of human factors assessments highlights their collective dimension, which excludes the canonical model presented in the general introduction. The series of rules that seem to govern the assessment process are more reminiscent of the organization of a production system than merely the implementation of a pre-established methodology that can be applied by a subject, as is the case in occupational medicine, for example[148].

Furthermore, some of these rules seem to organize the principle of open debate, since debates with the licensee are anticipated. Assuming that the principles of independence and transparency are equally upheld, the organization of human factors assessments could bring these assessments closer to the procedural model.

This organization of the assessment does not seem to be without consequences on the skills required to perform human factors assessments. In addition to having academic knowledge in their field, human factors experts will doubtless have to be proficient in certain skills specific to the production of nuclear safety assessments. For example, they will have to be able to interact with experts from so-called technical fields, licensees and the representatives of the ASN, as stipulated by the assessment rules. It is possible that they will have to take into account the particularities of France's nuclear safety institutions in order to do so. In such case, human factors assessments would be determined by these particularities.

---

[148] Dodier, N. (1993). *L'expertise médicale. Essai de sociologie sur l'exercice du jugement.* Paris, Métailié.

Conclusion to Part One: Historical and Institutional Influences of Human Factors Assessments?

This historical review of nuclear safety institutions and principles and of the incorporation of human factors in the production of assessments is ripe with lessons. In the span of fifty years, the role of the nuclear safety expert and the conditions under which they carry out their work have changed considerably. The physicist who was involved in the start-up and operation of new reactors gradually morphed into the expert statutorily separate from licensees, whose activity could be guided by a few key policy principles and governed by a set of procedures. Let us review three points that seem especially important:

- The resonance of defense in depth and the concept of barriers in nuclear safety policy principles. *It bespeaks a view of safety as a set of systems that prevent accidents or mitigate their effects. An indication of this is found in the prevalent form of control employed by human factors experts. Viewed in terms of human and organizational factors, such an approach to safety might lead an expert to check for the presence of operating training systems or rules on the design of operating instructions. The control of safety by experts could thus come primarily under a form of bureaucratic control as defined by Ouchi (procedure-based control, cf. p. 20).*

- Institutions that have historically exposed themselves to much criticism about their opacity. *Admittedly, recent internal changes have made the roles of institutions clearer to civil society. However, in technical dialogue institutions, described by some as "French cooking", one can expect to encounter expert capture phenomena favored by certain persisting national variables, i.e. the limited number of licensees*[149]*, the excessively close ties between experts and utility directors who attended the same schools (whether employed by a utility or an institute, human factors experts are graduates of the same universities);*

- The troubled incorporation of human factors in the assessment institute. *We have seen that human factors assessments necessitated access to actual facilities and that this access had to be negotiated. The difficulties in obtaining this access contributed to the failure of a project to build up the knowledge base of a relatively unknown field. Despite this setback, human factors were incorporated in nuclear safety assessments. This paradox strengthens the relevance of our initial questions: What knowledge is required to conduct human factors assessments? How, under these circumstances, can an effective assessment be made? Have management systems for building and sharing knowledge been set up? The beginnings of an answer are provided by the inventory of objects and topics assessed, which were gradually extended to include organizational aspects.*

This historical overview will be followed by a presentation of the current human factors analysis department at the IRSN as well as the organization of human factors assessments, which underscores the necessity to coordinate assessment stakeholders (the IRSN's specialist and generalist experts, ASN representatives, licensees). These procedures making up the assessment move it away from the canonical model and might draw it closer to the procedural model. It is an assumption that must be tested using our empirical data.

---

[149] This factor was identified in particular by Ayres, I. and J. Braithwaite (1992). *Responsive regulation. Transcending the deregulation debate.* New York, Oxford University Press (cf. pp. 54–55).

Part Two.
The Assessment Factory

The list of the various stages culminating in the formulation of recommendations prompts a view of human factors assessments as a production process. And yet this list and the presentation of the organizational conditions of the assessment and its procedures seem insufficient to adequately describe such a process. Indeed, they reveal nothing about the nature of the knowledge used by human factors experts, or about the nature of the relations among the multiple parties involved, or about the nature of the decisions and choices leading to the assessment's findings. Scope definition, review, analysis, drafting, proofreading, validation, integration of a contribution, explanation, comparison with the licensee, review by representatives of the nuclear safety authority, are stages in the assessment that must be described in detail in order to explain the process behind the production of an assessment report. We therefore opted for a material that seemed better suited than "semi-managerial" interviews. We therefore, with their consent, studied the human factors experts at work.

The first stage consisted in selecting the cases to be analyzed. Once we became familiar with this activity, we closely studied several assessments. Three of the cases followed made up the core material of the analysis and will be explained in this part.

1. SELECT

The choices made to select then study the assessments illustrate remarkably well what Jacques Girin called "methodical expediency"[150]. Indeed, from the very start the process of selecting the sample depended on the cases handled by the department's experts. Owing to the vast number of cases under way, we defined a set of characteristic values in order to select the ones we would study (cf. Table 4).

Variable	Possible values of the variable
Licensee type	CEA, EDF, Areva
Facility type	Power reactor (nuclear power plant), experimental reactor, R&D laboratory, nuclear fuel cycle plant
Assessment type	Incident analysis, contribution to a safety review, cross-disciplinary assessment
Role of the HF experts	Contributor, Coordinator
Examination by the advisory committee	Yes, No

Table 4: Characteristic variables of the cases.

We also split the case production process into the following stages:

- *scope definition stage (establishment of the assessment's scope and objectives);*

- *review stage (data collection);*

- *drafting stage (production of the report by the human factors experts);*

- *transmission stage (transmission of the report to the assessment's "clients");*

- *post-transmission stage (effects of the assessment after its transmission).*

This breakdown was not intended to be exact. Rather, its purpose was more to develop a tool to assist in the selection of cases that would provide enough data to cover the entire assessment production process.

The variables in Table 4 and the stages identified were used to produce a representative sample of the diversity of the assessment activity of human factors experts (cf. Table 5)[151]. Due to schedule conflicts, we were unable to study any of the assessments conducted in the fuel cycle facilities operated by Areva. However, every other case type defined using the variables explained above is represented in the selected sample.

---

[150] Girin, J. (1989). L'opportunisme méthodique dans les recherches sur la gestion des organisations.

[151] For ethical reasons, the names of the facilities have been changed.

	Operating organization	Blumenau incident	Minotaure review	Skills management	Artémis incidents	Artémis review	Safety mgmt.
Licensee type	EDF	EDF	CEA	EDF	CEA	CEA	EDF
Facility type	Nuclear power plant	Nuclear power plant	Experimental reactor.	Nuclear power plant	R&D laboratory	R&D laboratory	Nuclear power plant
Assessment type	Cross-disciplinary assessment	Incident analysis	Safety review	Cross-disciplinary assessment	Incident analysis	Safety review	Cross-disciplinary assessment
Role of the HF experts	Contribution	Contribution	Contribution	Coordinator	Contribution	Contribution	Coordinator
Examination by the advisory committee	Yes	No	Possible	Yes	No	Possible	Yes
Assessment stages followed	Transmission, post-transmission	Drafting, transmission	Review, drafting, transmission, post-transmission	Review, drafting, transmission, post-transmission	Scope definition, review, drafting, transmission, post-transmission	Scope definition, review, drafting, transmission	Scope definition

Table 5: The assessments studied.

In March 2005 the cases to be followed were selected and the data collection stage could begin.

2. FOLLOW

This stage of our research covered seven cases and lasted for nearly two years. We did not follow all seven assessments in the same way; we were involved more in the production process for some than for others.

In the case of the "operating organization" and "Blumenau incident" assessments, which were "no-load tests" that we chose not to explain, we merely stayed abreast of the evolution of the case and attended a few assessment presentation meetings (as a member of the SEFH). After the operating organization assessment was completed, we examined the assessment's consequences by reading letters sent by the licensee and by talking with the human factors experts and a representative of the ASN.

The bulk of our material consisted in studying the "Minotaure review", "skills management" and "Artémis incidents" assessments. Once the assessments were started (Artémis incidents) or not long after their scope was defined (Minotaure review and skills management) we went with the experts to coordination meetings and technical meetings held with the licensee's representatives. We were introduced as a PhD student at these meetings. As agreed, we maintained a discreet and silent presence during the technical discussions. After each meeting we sought the viewpoints of the experts and of the line managers and generalist engineers. We had access to all documents sent by the licensee and to the successive versions of the draft assessments. Once the assessment was submitted, we interviewed the various "clients" of the assessment (generalist engineers of the IRSN, ASN representatives, members of the advisory committee) and representatives of the licensee in order to obtain their views about the work of the human factors experts. In the case of the "Minotaure review" assessment, we stayed relatively in the background of the expert's work. However, we were included as a member of the "project team" by the expert in charge of the "skills management" assessment. The expert in charge of the "Artémis incidents" assessment found our interviews to be highly valuable. It is no overstatement to describe them as working meetings that allowed him to gain a better understanding of the incidents. Altogether, we attended forty-four meetings and conducted eighty-four interviews for these three cases.

We made the leap for the Artémis review assessment when we were one of the two experts in charge of the human factors contribution. A full-fledged participant in the assessment, we directed questions to the licensee's representatives and wrote a small portion of the contribution. As the safety review is illustrated by the Minotaure facility and incident analysis by the R&D laboratory, we will not go into detail about the process of this assessment[152]. Lastly, we participated in the meetings to define the scope of the "safety management" assessment.

Over the course of these two years, we amassed a relatively great deal of material. We made it a point to take many notes both during the meetings we attended and our interviews. The pragmatic approach adopted was unable to limit our research to an *a priori* wholly defined collection of data. So then how were we to cope with the "infinity of the sensory world"[153]? and its pitfalls – "when one does not see what one does not see, one does not even see what one is blind"[154]? Through what operations do we plan those parts of life to which we have access on the pages of our notebooks?

In order to be able to explain the assessment, it was necessary that the investigation make it possible to reconstruct the dynamic used by the experts to form their opinions. During the meetings and interviews, we therefore focused our attention primarily on the reasons behind differences of opinion, controversial issues and agreements as well as on the lines of reasoning and

---

[152] Our immersion in the assessment process was nevertheless highly informative as it allowed us to see first-hand the constraints placed on the expert.

[153] Aron, R. (1969). *La philosophie critique de l'histoire.* Paris, Librairie philosophique J. Vrin, p. 220.

[154] Veyne, P. (1983). *Les Grecs ont-ils cru à leurs mythes?* Paris, Seuil, p. 127.

rationale used during discussions. This is in particular what Jean-Yves Trépos suggests when, while creating his sociology of assessment, attached "great importance to argumentations, if by that one means the arrangement of persons and objects that allow the expert, sponsor and recipient of his decision to come to an agreement."[155] Furthermore, we read the assessment literature while studying the cases. We quickly contemplated comparing our empirical data with the assessment models presented in the general introduction[156]. All while striving to remain inquisitive in the face of the facts and empirical, as defined by Jean Wahl, this choice allowed us to guide our observations.

3. EXPLAIN

The following three chapters are devoted to explaining the data we collected while observing the assessments. The first discusses the contribution of human factors to the safety review of the Minotaure experimental reactor. The second looks at the analysis of two incidents that occurred at the Artémis research and development laboratory. The third focuses on the management of operating personnel at nuclear power plants.

Each chapter is structured in the same way[157]. Each begins with a brief introduction of the protagonists directly involved in the assessments and the facilities[158]. Then the assessment process is described gradually, stage by stage[159]. We will thus attempt to explain how assessments are constructed, the argumentations and recommendations of the human factors experts", and the nature of the relations among the people involved in these processes. At the end of each case we will summarize the key operations of the assessment, which we will discuss by referring to the models and theories presented in the general introduction.

---

[155] Trépos, J.-Y. (1996). *La sociologie de l'expertise*. Paris, Presses universitaires de France (coll. Que sais-je ?). 58.

[156] However, it was not until we finished collecting the data that we decided to use Ouchi's theory and the capture theory.

[157] These chapters are based primarily on information contained in our notes and in documents provided by the licensee and exchanged among the experts (requests, reports, presentations, questionnaires). We also developed a distinctive iconography, inspired by the supporters of the pragma-dialectical theory of argumentation, to explain the changes in the successive versions of the assessments during the drafting and transmission stages van_Eemeren, F. and R. Grootendorst (1996). *La nouvelle dialectique*. Paris, Editions Kimé, van_Eemeren, F. H. and P. Houtlosser (2004). Une vue synoptique de l'approche pragma-dialectique. *L'argumentation aujourd'hui. Positions théoriques en confrontation*. M. Doury and S. Moirand. Paris, Presses Sorbonne Nouvelle: 45–75.

[158] We have withheld the names and sex of the protagonists in order to protect their identities.

[159] The post-transmission stage will be discussed in part three.

Chapter 3. Contribution to the Minotaure Safety Review

"Mistrust is the mother of safety."
Jean de Lafontaine

In early 2000 the ASN sent a first clear-cut and definite signal in a letter announcing a safety review at the Minotaure facility (cf. Sidebar 4). The letter, dated February 16, was addressed to the licensee and stated that in order to be able to renovate Minotaure so that it could continue operating, the CEA would have to perform a safety reassessment. The letter also included a list of documents[160] that would have to be sent to the ANS. Once the necessary documents were sent to the ASN, the IRSN's experts would have to organize the safety assessment.

The ASN received a portion of the necessary documents in September 2001. These documents were reviewed by the IRSN's generalist engineers charged with assessing safety at the Minotaure facility to determine their eligibility. Their findings were incorporated in a letter sent by the ASN to the licensee on May 16, 2002. The following point in the letter is relevant to our discussion: "You informed me of your intention to analyze the risks related to human factors after the facility is renovated. However, I ask that you take these risks into account as soon as you begin your studies of the I&C renovation and handling operations."

Two assessments were planned. The first would be held to allow the ASN's representatives to decide on the "safety options" presented by the CEA to renovate the Minotaure facility and the safety review principles used. The

---

[160] "A document explaining the experience gained since the previous safety reassessment (OEF); an updated version of the safety analysis report (SAR), technical requirements (TR) and general facility operating rules (GOR); a study of the relevance of the design-basis accident; and an updated version of the earthquake-resistance study of the facility."

second, to be held in 2008 or 2009 after completion of the renovation, would allow the ASN's representatives to decide on the safety of the renovated reactor before restarting it.

All of the documents requested by the ASN in order to begin the first assessment were received in late 2004. The IRSN's team of generalist engineers in charge of Minotaure (coordinator, unit head and department head) was then brought in. The assessment was led by the generalist engineer in charge of coordinating the review of the Minotaure facility. His supervisor, the unit head, contacted his specialist counterparts in order to make sure they would take part in the assessment.

> 1) Geology – Liquefaction – Flooding
> 2) Seismology + Soil spectra
> 3) Earthquakes – Review methodology
> 4) Civil engineering – Earthquakes,
> 5) Earthquakes – Equipment resistance
> 6) Electric power supplies + Instrumentation & Control
> 7) Computer systems
> 8) Fire/Explosion
> 9) Radiological protection/Dosimetry
> 10) Criticality risk
> 11) *Human factor*

Sidebar 3: List of specialties involved in the Minotaure assessment
(source: IRSN document).

As shown in Sidebar 3, human factors were included. An expert was appointed to draw up a contribution. This expert had been conducting human factors assessments for the past eight years and was accustomed to dealing with cases related to the CEA's facilities. However, this was his first time on the Minotaure reactor and this was the first time a human factors assessment would be conducted at the facility.

> The Minotaure reactor is dedicated to determining neutronic characteristics used for sodium-cooled fast reactors. Making assemblies from very long standard rods (3.80 meters) containing small individual elements – strips, platelets made from highly diverse materials (uranium, plutonium, sodium fuels, steel, absorbents, diluents, etc.) – makes it possible to recreate any type of core with a high degree of accuracy and thus similarly vary the parameters studied. Fifty different cores (requiring the handling of 25,000 rods, or around 45 million strips or platelets in the case of major programs) were created between the reactor's commissioning in the 1960s and the first safety review in 1988.

Sidebar 4: What is Minotaure? (source: IRSN documents).

1. THE SCOPE DEFINITION STAGE (FEBRUARY 2004–MARCH 2005)

By "scope definition", we mean the stage during which the topics to be assessed are selected. Initial contacts are established between the human factors expert and the assessment coordinator, either bilaterally or during group meetings set out in the institute's quality procedures. The human factors expert reads the file prepared by the licensee then goes to the facility to take part in a tour and attend a general presentation. These various steps allow the expert to draw up an initial list of topics that will serve as a basis for his investigation.

The assessment coordinator and the human factors expert first met in December 2004, at which time the expert informed the coordinator of his desire to meet the facility's operating personnel. A summary of their discussion is provided in Sidebar 5.

The construction of the contribution of the human factors experts is a group effort led by the generalist engineers. They organize meetings that can have a strong impact on the work of the contributing experts. As a result, they are pivotal components of the production system. During the scope definition stage, the following three meetings took place on the IRSN's premises:

- *meeting prior to the kick-off meeting;*
- *kick-off meeting;*
- *internal consultation meeting.*

Held on January 12, 2005, the *meeting prior to the kick-off meeting* was attended by four experts and six representatives of the licensee. Four of the licensee's representatives whose jobs were directly related to the operation of the facility were present (the facility manager, his safety assistant, a safety engineer and the manager of the facility renovation project). The meeting was chaired by the three generalist engineers (the assessment coordinator, the unit head and the department head). According to the minutes of this meeting, "regarding the request of the human factors experts to interview the facility's operators, the IRSN committed itself to provide the CEA with detailed information on the purpose and conduct of these interviews when such will be concretely considered."

One week later, three representatives of the ASN, a higher number of CEA representatives, the director of the IRSN's reactor safety division and a few IRSN experts took part in the assessment *kick-off meeting*.

Characteristics of the Minotaure reactor

"Minotaure is characterized by a high human contribution to safety on account of the many manual operations required to run the facility and which may lead to risks of criticality, radiation exposure and contamination. These human activities are:

– filling ... of the various rod types [containing platelets of uranium, plutonium, sodium fuel, steel, absorbents, diluents] ... by hand at dedicated workstations;

– movement and transfer operations, either in the storage rooms or the reactor core, using special handling equipment ...; transfer operations of simulation elements in the fueling room (SCM) and the various stores;

– operation of the reactor;

– core fueling/emptying operations; control rod or mechanism loading/unloading operations."

Objectives of the human factors review

The human factors experts will have to decide on "the inclusion of the human factor in the safety review approach." The review will also cover "the guidelines on the planned changes to the facility", and particularly on the "safety options relating to the I&C renovation".

In addition, the human factors expert is asked to give his or her opinion on "the lessons learned from the incidents and anomalies with a significant human component" and "the analysis made by the licensee relating to the human and organizational factors in its facility."

Conditions of the human factors review

In addition to technical meetings, the human factors expert would like to interview the operating personnel.

Sidebar 5: The first exchanges between the assessment leader and the human factors expert.

The general technical aspects of the facility (the technical characteristics of the various buildings and the reactor power, amongst others) were discussed during the *internal consultation meeting*[161] held on February 1. The meeting was held by the generalist engineers and attended by most of the contributing experts. For the latter, the meeting served in a way as an opportunity to get up to speed. It also allowed them to meet each other and discuss the interfaces. During the meeting, the human factors expert identified an interface with the specialized area of criticality. The two experts would work together to assess the measures implemented to control criticality risks related to fuel rod fabrication operations (cf. Sidebar 6).

---

[161] According to the IRSN's quality manual, the internal consultation meeting should have preceded the kick-off meeting.

Fissile materials handled outside nuclear reactors, i.e. in laboratories and plants and during transportation operations, represent a particular risk known as the criticality risk. This is the risk of creating the conditions necessary to set off and sustain chain reactions. For example, this risk arises when enriched uranium with a concentration of 3.5% 235U is handled in quantities in excess of 60 kg.

Criticality risk prevention imposes taking special measures that are analyzed and studies at all operating stages. It involves, for example:

– observing, with a sufficient safety margin, a limit on the mass of fissile material under which chain reactions become physically impossible.

– using "safe geometry" equipment, i.e. equipment that is sized to render physically impossible the initiation of a chain reaction, based on the mass of the fissile material present.

At Minotaure, the criticality risk is particularly present when operators handle fissile materials during fuel rod fabrication.

Sidebar 6: Criticality risk (source: IRSN documents).

This internal consultation meeting between generalist and contributing experts took place prior to a tour of the facility. The tour, held on February 2, was organized by the licensee and attended by the human factors expert.

Although we have emphasized the group aspect of the process, there were times when the expert worked alone. From the moment he received the documents from the coordinator, he "conducted the review" in a more traditional sense of the term. However, few documents were of direct relevance to the human factors expert, and those stamped as such were even fewer in number. The safety analysis report, which listed all of risks identified in the facility and categorized them by specialized area, did not contain a chapter on human-factor-related risks. "It was the same for the GOR[162]. They revealed little about the human factor" (the human factors expert, interview of February 22, 2006). The matter of human factors was also absent from the facility's annual safety records and significant incident reports. On February 17 the coordinator faxed a request for ten additional documents. These documents included memos on the organization of the laboratory operating the facility and its overseeing department, quality-related documents, shift supervisor training procedures, and a document on the receipt and handling of a specific type of platelet. The coordinator also provided the experts with the list of twenty-nine significant incidents that had occurred at the facility since 1989.

---

[162] "The general operating rules (GOR) present the measures implemented during the operation of a reactor. They round out the safety analysis report, which essentially deals with measures taken from the moment the reactor is designed." (source: SFEN document)

At this stage of the assessment, and based on the data in our possession, the relations between the various people revolving around the human factors expert and their roles may be described as follows:

- The coordinating expert *was the expert's main contact. He was, in a manner of speaking, a project manager. He provided the human factors expert with necessary information (characteristics and history of the facility); expressed his expectations, particularly in the request, thus making a strong contribution to the formulation of the topics to be assessed; and discussed the submission date of the contribution with the expert.*

- The coordinator and human factors expert's supervisors *consulted each other before assigning the human factors expert to the assessment (need for a balance between supply and demand). The head of the SEFH was brought in by the coordinator by means of the request. He was rarely called on by the expert at this point in the assessment.*

- The criticality expert. *The human factors expert identified the need for collaborating with the criticality expert. The coordinator subsequently organized their initial meetings.*

- The licensee and the ASN. *The coordinator placed the human factors expert in contact with the licensee and, if need be, the ASN. The expert met the facility's representatives during a tour. It is worth mentioning that the licensee asked the expert to justify his requests to interview members of the facility's operating personnel. Since the licensee had a say in how the expert conducted his assessments, the expert had to negotiate from the start. There was no direct contact between the human factors expert and the representatives of the ASN.*

The result of these interactions? An initial list of topics, associated with a review strategy, designed to allow data to be collected and argumentations to be developed in order to meet the expectations listed by the coordinator in the request. The topics and the data collection methods identified at this stage of the assessment are listed in Table 6.

Topics	Data collection methods
Consideration of human factors in: • the facility safety review; • the renovation of the control room; • the design of the remote shutdown panel; • maintenance operations; • safety-sensitive operations; • operating experience feedback.	• questionnaire and technical meeting with the licensee, in the coordinator's presence; • collaboration with criticality experts on safety-sensitive operations; • interviews of operating personnel.

Table 6: The outcomes of the scope definition stage.

2. THE REVIEW STAGE (MARCH–JULY 2005)

This stage of the assessment by the human factors expert was marked by the following key moments:

- *the preparation of a questionnaire to be completed by the licensee and discussed during a video conference;*

- *the holding of a progress meeting, which was a milestone of the review stage;*

- *the analysis of several documents, include a summary of the licensee's human factors analysis, the human factors recommendations" and the functional specifications for the renovation project, the significant incident reports and the facility's safety records;*

- *a second technical meeting with the licensee in the Minotaure facility, prepared beforehand with the IRSN's criticality experts.*

2.1. The Questionnaire and the First Technical Meeting

The human factors expert prepared his questionnaire in late March 2005 (cf. Sidebar 7). A few weeks later, the licensee sent back its answers along with a memo in which the facility manager stated that "the responses were written conjointly by the team in charge of renovating the I&C system and [a] human factors expert from the CEA."

They were discussed at the first technical meeting, a video conference held in early June. The meeting was attended by the assessment coordinator and the IRSN's human factors expert, the four facility representatives directly involved in the safety review (the facility manager, the safety assistant, the safety engineer and the renovation project manager) as well as a human factors expert from the CEA.

1) Safety review: how are human factors addressed in the safety review? What "regulatory, technical and normative standard" is used?

2) OEF: how is operating experience feedback collected and analyzed?

3) Renovation of the control room: how are human factors factored into the design project? Do the chosen arrangements take into account "human-error tolerance"?

4) Remote shutdown panel: how are human factors considered in the design of the remote shutdown panel?

5) Operational activities: what are the design requirements and how are the chosen arrangements validated? "(human-machine interaction, human-machine interface, workspace layout, workstation layout, organization of work, facility operation training)"

6) Maintenance activities: "Your records do not clearly indicate how human factors are integrated in maintenance operations. This should be specified."

7) Sensitive operations: how are safety-sensitive operations identified?

8) Manual override of automatic functions: is this matter addressed?

Sidebar 7: Summary of the questionnaire prepared by the human factors expert.

Allow us to dwell a moment on the data collected by the human factors expert at this point[163].

The licensee's written response indicated that its consideration of human factors for the safety review was predicated on "two complementary approaches":

– "An analysis of the general organizational arrangements at the facility (macroscopic approach) and of arrangements more specific to certain work activities identified as being sensitive to the safety of the facility (microscopic approach)[164]."

– "Sensitive operations were identified in collaboration with the facility and in particular with the facility manager, the safety personnel and the operations manager. The activity analyses, which were based on observations of real work situations, were used to identify how some factors of work situations lead (most

---

[163] The passages used by the human factors expert to support his findings are underlined on the following pages.

[164] Here is how sensitive operations are defined further on in the document: "Sensitive operations are 'defined as operations whose normal functioning is essential to maintaining safety; in other words, those operations which, if they malfunction, lead to immediate or latent degradation of safety .... Activities comprising sensitive operations meet the following criteria: be performed in major part by operators; have potentially highly severe consequences in the event of non-recovery from failure; be representative of the activities of the facility.'

often potentially) to human error. This in-situ approach to the activity was rounded out by the examination of the operating experience feedback and the interviews. General organizational arrangements and arrangements specific to sensitive activities will be implemented as a result of the analysis."

During the meeting, the human factors expert requested further details on the human factors analysis. The replies provided were scarcely more informative that those that had been given by fax. Nevertheless, the licensee made an effort to convince his counterpart of the seriousness of the study with replies such as "The analysis was made by a company you work with" and "The study lasted for more than a year" and "The meetings were attended by a human factors expert".

The reorganization of the facility, which was neither the subject of any written questions nor mentioned in the request, was discussed right at the start of the meeting. Here is an excerpt of the discussion:

– The licensee: Reorganizations are pending validation.

– Human factors expert: These reorganizations are going to have an impact on the activities. How was this impact analyzed?

– The licensee: The reorganization is headed in the right direction. Currently, we have two large experimental laboratories [Minotaure and another] that are headed by the same person, who is based at Minotaure. These laboratories are located far apart from each other and their activities are highly different. Minotaure is undergoing a safety review and being renovated. The other laboratory is at steady state. *The human factors analysis revealed weaknesses in this organization. To make this system more efficient, we have decided to appoint a separate head for each laboratory.* This way, all personnel on each site will report to one person. That's the foundation of safety. It's very important for routine operations.

Regarding operating experience feedback, the licensee explained in its written response that the situation of the facility since the previous review, held in 1988, is described in a document. This document "addresses the organizational measures as well as operator training and skills management measures implemented. It also discusses the incidents that have occurred at the facility. The lessons learned from these incidents have not been specifically analyzed in terms of human and organizational factors. However, these aspects underlie the analyses presented. ... At the end of this stage of the study the facility drafted a guide of design recommendations ... for the contractor in charge of renovating the facility's control room."

During the meeting, the human factors expert raised again the matter of *the absence of the formalization of operating experience feedback.* The licensee acknowledged this, but re-emphasized the quality of the work performed: "The incident analysis may not have necessarily been formalized, but many human factors elements were included in it."

The expert then asked *whether a human factors expert from the CEA participated in any incident analyses. The answer was no.*

Regarding the operating documentation, once again the licensee's written response was reassuring, but not very explicit: "It is quite clear that specific action will be devoted to [the documentation] .... Furthermore, the analysis of the study conducted during the safety review identified the necessity to start thinking about the operating documentation."

The matter of validation was also discussed during the meeting:

– Human factors expert: How will the operating documentation of the renovated facility be validated?

– The licensee: The contractor is supposed to give us lots of information. *We haven't yet defined the validation process*, but we've outlined it.

– Human factors expert: Will a human factors expert be present during the validation tests?

– The licensee: It's too early to tell. We'll get along as best we can. Having easy-to-read operating documents is very important. The human factors hub will make sure of that.

As will be seen, this response did not fully satisfy the expert, who would go on to emphasize validation aspects in his assessment.

Fuel rod fabrication was one of the sensitive activities discussed during the meeting:

– Human factors expert: Have falling rods been identified? *Could such a fall be caused by human error?*

– The licensee: No, that's never happened. The rods aren't moved manually by the operator. That's locked.

It was agreed that there would be tour of the facility to see the rod fabrication activities during a forthcoming technical meeting on the licensee's premises.

When questioned about the integration of human factors in the project to renovate the control room, the licensee answered that the project included a human factors approach. "The methodological requirements of human factors are covered by a specific chapter in the project specifications and apply to the overall design .... The project monitoring process includes a human factors component for which the facility is assisted by the CEA's human factors hub. The CEA sends written reports, comments, analyses and letters to the contractor so that human factors requirements are met."

When asked by the expert about the contractor's human factors skills, the licensee replied "The contractor wasn't up to scratch at the start. But it has since set up a team consisting of an engineer and a human factors expert with [*XXX*]."

Once again, validation aspects were discussed:

– Human factors expert: How are design choices validated?

– The licensee: The operators are involved in the iteration. Interactions between operators and designers seem to us to be the most efficient process with simulation scenarios. It's essential to validation and training.

At the end of meeting, which lasted for nearly two hours, the coordinator and the human factors expert shared their initial impressions:

– Human factors expert: They've provided lots of information. Their approach looks to be positive, especially in macro terms. However, I'm having a tough time reaching a decision in micro terms [sensitive activities]."

– Coordinator: They appear to have given the analysis proper consideration.

2.2. The Progress Meeting – a Review Milestone

The progress meeting was attended by thirteen representatives of the CEA, two representatives of the ASN and nineteen experts from the IRSN. It was held one week before the video conference.

Prior to this meeting, the human factors expert and the coordinator agreed that the expert's presence was not necessary and that the coordinator would present the initial findings of the analysis in his place. Every specialized area was reviewed during the meeting. When the meeting turned to human factors, the pilot noted the "very high human contribution to safety (many manual technical and handling operations that can lead to risks of criticality, external exposure or contamination)." He also pointed out that the analysis of the questionnaire results revealed that "the approach used to take human factors into consideration was satisfactory in principle." He concluded that "No hold points have been identified." These comments explain why the human factors expert and his supervisor were absent from the meeting.

During the meeting, the possibility of the human factors expert interviewing the facility's personnel became less likely. The coordinator related the situation to the human factors expert in an email: "The licensee is not against the principle of interviewing the facility's personnel. However, it wonders (as do some people [in my department] about the necessity of conducting such interviews at this stage of the review and, in particular, during [this first assessment] on the reassessment principles and methodologies as well as the facility renovation guidelines adopted. (...). It would be a good idea if you wrote back telling me of your view on the matter and in what way these interviews are part of the current review and how you plan on conducting them at the Minotaure facility."

The human factors expert understood the point. Considering that the interviews were not crucial, he dropped the matter and discussed it no further with his supervisor.

2.3. The Review of New Documents

Analysis of new documents provided by the licensee began a few days after the first technical meeting. These documents were the summary of the human factors analysis, excerpts from the functional specifications for the renovation project and a guide of human factors recommendations for design. The review of these documents contributed to identifying the following new topics:

- *Activity planning;*
- *Licensee/maintenance and licensee/researchers interfaces;*
- *Skills management;*
- *Sensitive activity: fuel rod handling.*

2.4. The Second Technical Meeting

At the end of the Q&A session on fuel rod fabrication operations during the first technical meeting, the facility manager proposed to take the human factors expert on a tour of the facility. The human factors expert accepted, asking that the facility's new organization be presented during the tour. He also asked to see the meeting minutes and documents drafted during the renovation project. These three items were put on the agenda of the second technical meeting held in the facility.

A detailed presentation was made of the facility's new organization ("one head per laboratory"). The human factors expert asked if there were any plans to *investigate the "potential risks of adverse effects related to the implementation of this new organization,* particularly since the facility's personnel will have to change long-established work habits." The facility manager replied that the proposal for a new organization had been discussed with personnel and that their feedback to date was by and large positive. He also explained that the new organization would result in more efficient feedback of problems encountered on the job: "I insisted that emphasis be placed on the 'operations manager/safety manager' relationship. It's clearer and more reliable." As will be seen later, the human factors expert would however consider that these risks would have to be assessed.

During the discussion the human factors expert gave his opinion on the summary of the human factors analysis he had recently received. "*It's a highly positive initiative, but the document does not list the facility's strengths and weaknesses in terms of human factors or contain observable facts of interest to human factors.* It does not really establish the link between the observable facts, the weaknesses and the recommendations." The licensee made no comments.

The human factors expert stated that he had read the functional specifications, pointing out that "*the various deliverables, acceptance tests or acceptance criteria are not listed in a part of the human factors specs.*" The licensee did not deny this but explained that "they are discussed and negotiated with the contractor as the project advances."

The afternoon was devoted to touring the loading room, where the fuel rods are fabricated. Earlier, the human factors expert met with the criticality risks expert to delve more deeply into this subject with the licensee. Together, they drew up a list of questions that they sent to the Minotaure team prior to the meeting to guide the discussions. A few extracts of the questionnaire are provided in Sidebar 8.

---

1) *Human error analysis*

The loading room can contain various fissile materials with different mass limits. Switching from one fissile material to another is a very significant source of error. Have you analyzed the potential human errors at each operating stage? What are the potential safety consequences in the event of human error during these stages? And during the rod disassembly stages?

2) *Organization of work*

Could you explain the organization used during tube design? (for instance, the number of operators required, the assignment of tasks regarding store management, removal from storage, material transfer, material counting and rod fabrication)

What measures are to be taken if an operator is absent?

According to the safety analysis report, the licensee implements a system of double-checks to ensure that criticality risk mass limits are not exceeded. Could you give us your current thoughts about the methods used to carry out these double checks? Could you explain the basis of the robustness of this system of double-checks to ensure against the risk of fitting more than the requisite number of strips and/or fissile or moderator grid straps?

3) *Training*

How many operators are trained in rod fabrication? How long have they been working at the facility? How does the licensee maintain and transfer specific skills? How does the licensee ensure that operators are informed about the risks associated with the handling of materials?

Sidebar 8: Questionnaire on the fuel rod fabrication activities (excerpts).

These questions were put to the safety engineer and the facility manager during a tour that was also attended by a criticality risks expert from the CEA. The tour was led by an operator whose duties included assisting in the fabrication of fuel rods. He reassuringly cited the technical, human and organizational countermeasures used for each disaster scenario brought up by the coordinator. The organization of rod fabrication tasks was discussed. A number of documents – in particular the rod sheet and the room requirements sheet – were presented. "The room requirements sheet is based on the rod sheet. It tells the store manager how many items to take from inventory based on the type of items needed for rod fabrication. *The management of material transfers by the manager is based on the number of items, not on the determination of the mass of the items to be transferred.*" This coordinator commented on this last point.

The licensee explained that a number of avenues for counting materials being removed from the store were being explored. He reassured the human factors expert that "We're considering adding, on the rod sheet, a check of the mass limits against the quantities of items required for each tube. Furthermore, the maximum numbers of elements that can be handled are being revised to cut back on the number of transfers made by the manager in order to reduce his radiation exposure. Once the right number of items is removed from the store, potential human errors that could occur during rod design have no impact on safety, just the quality of the experiments."

The team toured the store then went back to the loading room to see how the rod fabrication activities are organized. "At least three people, including the loading room manager and a store manager, must be present in the loading room during fuel rod fabrication." When questioned by the human factors expert, the licensee stated that each manager had a specific type of clearance but that *it was perfectly possible to imagine having just one manager with both types of clearance.* The tour then turned to the matter of the mass limits double-check. The operator explained that once the number of fissile materials was removed from the store by the store manager and placed on a cart, *the loading room manager would in turn count the number of fissile materials on the cart and make sure that it matched the quantity indicated for the rod* on the rod sheet. When questioned by the coordinator, the licensee replied that "given the criticality risk margins, a possible difference of one or two strips would be of no consequence." The loading room manager then moved toward the cart, which was by the loading table, and counted the non-fissile elements. They would also be placed near the loading table. The rod was now ready to be fabricated. The licensee explained that the elements were inserted in the rod by two operators, with each operator inserting one element: "They proceed by alternating a fissile element with a non-fissile element and they *check each insertion together.* The loading room manager ensures that each fuel rod is properly fabricated."

The tour lasted for more than two hours. At day's end, the human factors expert was satisfied. However, some issues – particularly those relating to skills management and training – had not been addressed. We will see later on how this deficiency was remedied.

A few weeks after the tour, the human factors expert drafted his minutes of the two meetings. While writing them, he kept in mind the wording of his assessment so that the evidence he would use to support his opinions would be clearly apparent. The coordinator proofed and edited the expert's draft and then sent it to the facility manager a few days later, on July 19. As no word was received from the facility manager, the draft became the final minutes of the two meetings.

2.5. Conclusion: Nature and Outcomes of the Interactions

This stage of the review was thus characterized by a revision of the list of topics drawn up during the scope definition stage and the identification of problems that would be used as evidence to support future opinions. These issues were identified and this evidence was collected through a documentary review process and interactions with the licensee. How may they be described?

First of all, it should be noted that the human factors expert appreciated the fact that a human factors expert from the CEA was present at the meetings. Both experts knew[165] and understood each other. The participation of a human factors expert bespoke a determination to include human factors right from the design stage. The licensee strove to convey this message to the expert from the IRSN.

The interaction between the expert from the IRSN and the licensee was often characterized by the licensee's efforts of persuasion. The wary expert ("What if ...?"), unfamiliar with the facility's routine operations and particularities, was at times bewildered by the licensee's upbeat and reassuring attitude ("Everything is under control"). Whenever it was hard for him to form a solid opinion, he relied on the immutable building blocks of the policy principles of proper consideration of human factors ("the validation conditions must be defined"; "operating experience feedback must be documented").

Relations between the experts (coordinator and human factors experts) and the licensee were courteous. Neither side ever raised their voice or made any demands or threats. The expert was there to collect data and give his opinion and the licensee was there to present safety and human factors at the facility and convince the expert that these issues were properly taken into account.

---

[165] The CEA's human factors expert was a former expert from the SEFH.

The meeting minutes allowed the expert to inform the licensee of his impressions and check whether the evidence on which they were based was "validated" by the licensee. This approach follows one of the principles of the IRSN assessment: "At the end of the review, the expert and the licensee must agree on the points of disagreement."

Among the resources available to the expert during the review stage, mention must be made of the assessment coordinator and his knowledge. The coordinator was the expert's first "client" and assisted and guided him during his investigation. He was by his side during the technical meetings. He asked questions and restated the expert's own questions in terms the licensee could better understand. He proofread the questionnaires and minutes and clarified certain technical that were beyond the human factors expert's grasp.

Topics	Issues identified
Reorganization	– Potential adverse risks not analyzed – Weaknesses of the human factors analysis, which formed the partial basis of the reorganization project
Operating experience feedback	– Not documented – No involvement of the CEA's human factors experts in the analysis of incidents and anomalies
Operating documentation	– Validation conditions not defined
Rod fabrication	– Possibility of failure to perform double-check during rod fabrication – Criticality risk caused by a counting error inadequately addressed
Renovation project	– Conditions for validation of design choices not defined – Deliverables and acceptance absent from the human factors specifications

Table 7: The outcomes of the review: the problems identified.

Lastly, the human factors expert collaborated with the criticality expert. Although we have little record of his influence, he did participate in creating the preparatory questionnaire for the second meeting. The criticality expert was particularly inquisitive about the robustness of the double-check, as he stated after his assessment: "We wanted to know to what degree a procedure could provide a strong line of defense for the criticality risk. The human factors expert brought out a list of conditions for an effective double-check. Many were from the aeronautics sector. That confirmed our gut feeling." (interview of April 13, 2006)

The outcomes of the various interactions the human factors had during the review stage are listed in Table 7.

Lastly, there were no interactions between the human factors expert and his supervisor. This would impact the drafting of the assessment.

3. THE DRAFTING STAGE (JULY–DECEMBER 2005)

One might think that during this stage the human factors expert worked primarily alone on his report. But as will be seen, his supervisor did more than just sign off on the report and considered that the review inadequately addressed a number of topics. An additional review would be conducted at his request. Furthermore, a new document provided to the criticality expert in early August would cause the human factors expert to change his opinion about the fuel rod fabrication activities.

In this section, we will present the supervisor's general critique of the human factors expert's first draft and the effects his comments had on each issue. We will describe in detail the proofreading process because, it seems to us, it clearly shows two different conceptions of the human factors assessment and, to a greater extent, two contrasting methods of risk management.

3.1. The General Comments of the Human Factors Expert's Supervisor

The expert submitted his report to his supervisor in mid-August 2005. A few weeks later, the supervisor sent it back copiously annotated and verbally told the expert what changes were necessary. In an interview, the expert's supervisor explained that "The topics are not sufficiently structured. There are many parts where I don't know if the opinion is based on the CEA's analysis or if it was given by the IRSN. An expert must draw a clear distinction between what he thinks and what the licensee shows. The report lacks descriptions about the licensee's operations, such as organization, skills management and rod fabrication. Secondly, the expert places too much importance on the CEA's analysis, which he critiques extensively. Also, he gives his opinion about the analysis. What I want is for him to give his own analysis of safety and the consideration of human factors in the facility. Thirdly, the expert's critique is not sufficiently developed at several points. Lastly, it lacks conclusions. An expert's opinions must have conclusions. If an expert has nothing to say about a subject he is addressing, he must say so. If he makes a number of critiques, they must lead to a request." (interview of September 6, 2005)

3.2. The Effect of the Supervisor's Comments on Each Topic

The supervisor and the expert discussed the various topics of the human factors assessment. The arguments used by the expert to support his findings for each topic are shown in diagrams on the following pages. His initial arguments are preceded by the letter "A" while those resulting from the discussions with his supervisor are preceded by the letter "B". The topic of discussion appears in the center of each diagram. Positive arguments about actions taken by the licensee appear above each topic and are denoted by a plus (+) sign. Negative arguments appear below each topic and are denoted by a minus (−) sign. A topic surrounded by a black border was the subject of a request. Lastly, arguments whose validity or relevance were opposed by the supervisor and were withdrawn by the human factors expert are indicated by struck-out text.

The topic of reorganization of the facility was reviewed by the human factors expert during the two technical meetings. The expert considered that, given the potential risks posed by a new organization and in accordance with the principles of continual improvement, the licensee would have to present a review of this new organization. He also considered that situation prior to the reorganization had been only partially analyzed, particularly since "organizational analysis issues were not addressed".

For the department head, it was necessary to list these issues and explain in what way they were important to the safety of the facility. When the expert stated that the reorganization could create siloing, his supervisor asked him to specify the reasons why this siloing could occur and explain the related risks. Faced with the difficulty of the task, the expert did not provide any reasons.

Although the supervisor's oft-repeated question regarding the request to perform a review, "do you have any particular apprehensions that justify your request" remained unanswered, the expert maintained it.

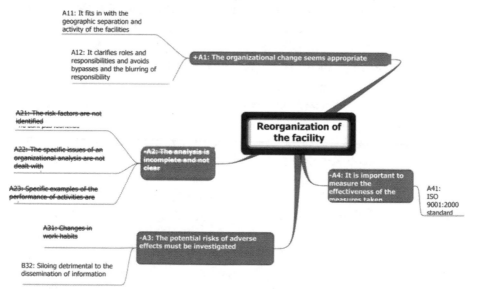

Identical diagrams on the development of argumentations and conclusions for the three areas of activity planning, operations-maintenance interface and operations-researcher interface are provided on the following pages.

The initial version contained only one negative critique, which did not result in a request. Following his discussions with his supervisor, the expert justified (mention of an incident in the case of activity planning) or withdrew (the interfaces) his critiques and added positive aspects to qualify his arguments and justify the absence of a request. The expert occasionally regretted this change, saying that "you're made to say the opposite of what you initially stated ...." (interview of March 9, 2006).

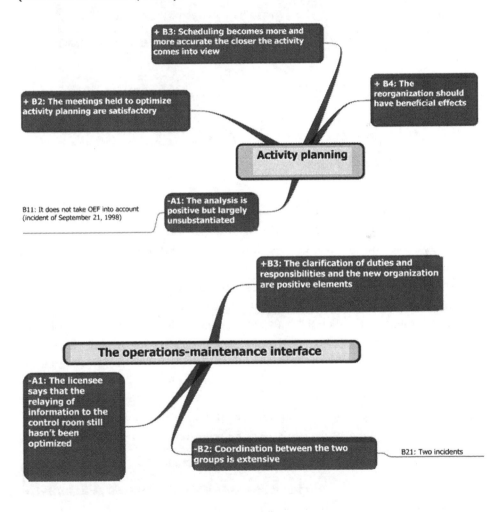

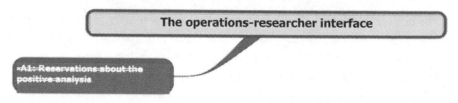

The topic of skills management was not discussed during the technical meetings. Most of the data used by the expert was derived from the human factors analysis.

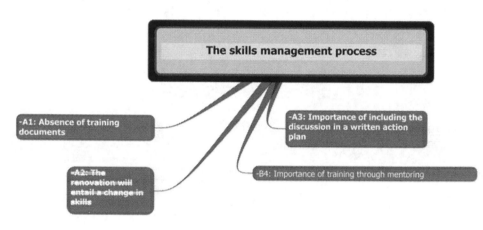

Since mentoring is important, the supervisor asked the expert to obtain information about the tutors, such as whether the rate of departure was compatible with the retention of resources. The expert added this item to the request he would make.

The topic of operating documentation had been reviewed during the technical meetings and the discussion of the human factors analysis. The licensee stated that thought was being given to the validation of the documentation and that the clarity of the instructions was a very important aspect. As a result, it said that it would call on the CEA's human factors experts. As validation was a crucial stage for the expert, he issued a validation request with his supervisor's assistance.

The topic of operating experience feedback had been reviewed in part during the technical meetings. The supervisor rejected some of the expert's critiques, describing as "unacceptable" his argument that "the statistics from other facilities are different from yours. Your methodology is therefore flawed". The supervisor asked the expert to draw on the incident reports, which he did. The supervisor also asked the expert to explain what he was expecting from the licensee when he wrote that "at the least, documenting OEF in the area of human factors is clearly indispensable." The human factors expert's response was to withdraw the sentence.

These arguments warranted the human factors expert's request to include a human factors chapter in the annual safety report. His supervisor fleshed out his request and suggested an additional request for an improvement in the analysis of incidents and accidents.

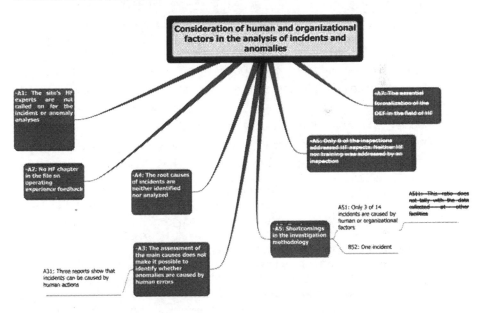

The bulk of the arguments on the operations activity came from the human factors analysis. The expert's supervisor asked him to show more specifically how operations were safety-sensitive. The expert did so by providing a relatively detailed description of potential initiating events caused by human error.

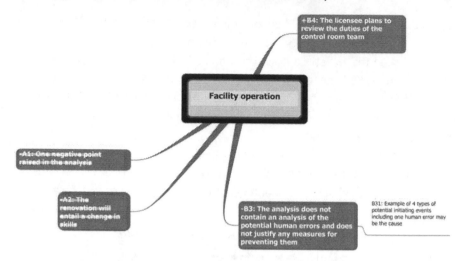

In the midst of the drafting stage, the direction of the assessment of the rod fabrication activity was suddenly changed by a new document from the licensee. This document, whose origin and course is worth pointing out, would result in qualifying the initial assessment. The sidebar below contains excerpts of correspondence between the heads of the IRSN's generalist engineering and criticality departments and the licensee. The tone of these exchanges is in stark contrast to the courteous relations described so far (cf. Sidebar 9).

> "Technical discussions held during a video conference on May 23, 2005, revealed a certain number of differences of opinion, the main ones of which were presented during the progress meeting held last May 30. After this meeting, you informed us by telephone of your disapproval of how the IRSN presented your criticality risk analysis explained in the documents provided. ... This letter is to remind you of the current difficulties resulting from the IRSN's review of your criticality risk reassessment case.
>
> You state that you are currently preparing a procedure that will specify "the organizational measures for conducting and documenting these checks". At present, the IRSN has no guarantee of the possibility of excluding the criticality risk in a manner consistent with FSR[166] I.3.c solely on the basis of organizational measures. We remind you that FSR I.3.c specifies "two independent failures" that must be of "low probability". In principle, this would be impossible with checks inevitably subject to human failure.
>
> Under these circumstances, and in order to allow the advisory committee to determine whether the objective set by FSR I.3.c would

---

[166] Fundamental safety rules (FSR) are ASN recommendations that define safety objectives and describe the practices that it deems satisfactory for achieving these objectives. There are some forty fundamental safety rules (sources: IRSN documents).

be "achievable", you should provide the following as quickly as possible:

    – more information on the organizational measures and the confidence that may be placed in them (e.g. in terms of line of defense),

    – any additional measures for meeting the overall objective of FSR I.3.c.

    Lastly, we draw your attention to the fact that whenever it is not demonstrated that measures meet FSR I.3.c, it is customary to choose authorized mass limits for fissile materials that are well below the acceptable limits. Such reduced limits case could severely constrain facility operation."

Sidebar 9: Excerpt of a letter signed by the supervisors of the generalist and criticality engineers and sent to the licensee on June 23.

A technical meeting between the licensee and the criticality experts was held on July 5. The licensee would refer to it in a faxed reply to the head of the generalist engineering department. An excerpt of this reply is provided in Sidebar 10 below.

"The July 5 meeting at the Minotaure facility provided the opportunity to constructively discuss the matters mentioned in your letter. It allowed the experts in charge of your case to tour Minotaure in order to grasp the particulars of the activities and operations carried out there. Regarding the planned measures for the facility in order to meet the general principles set out in FSR I.3.c, the organization of the operations performed (in the loading room) and the rules adopted were presented by the licensee during the technical meeting. This made it possible to focus on identifying sensitive operations and thus specify how the checks are to be conducted (criticality units concerned; parties, means). For this purpose, the licensee has proposed improvements and presented additional indicative calculation elements for evaluating the limit margins and thus the real risk."

Sidebar 10: Excerpt of the licensee's reply (August 2, 2005).

And this is how an "analysis of potential human errors and associated lines of defense" landed on the desk of the head of the generalist engineering department. It detailed seven human-factor-sensitive activities involved in the fuel rod fabrication process and listed the potential human errors, preventive measures and detection/monitoring measures for each. Although satisfied with the licensee's effort, the human factors expert considered that the analysis was incomplete and voiced misgivings about the effectiveness of the preventive measures.

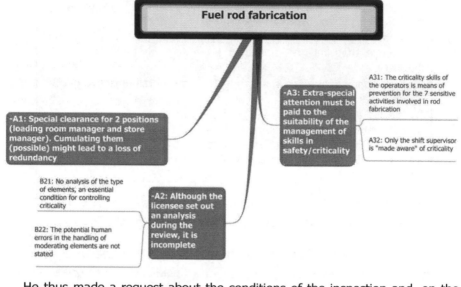

He thus made a request about the conditions of the inspection and, on the advice of his supervisor, another about training. He made a third request about the management of clearances when his supervisor deemed "inacceptable" that the employees could have several clearances. But, as will be seen during the transmission stage, the generalist engineers were able to dissuade him.

The human factors expert made a last request regarding the validation methods for the design choices for the I&C system, emphasizing the need to involve human factors experts in the validation process. This last recommendation would be struck out by the expert's supervisor, who explained that "we must judge the results, not the means."

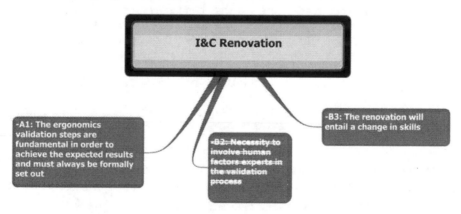

3.3. Conclusion: nature and outcomes of the interactions

The diagrams used to depict the interactions between the human factors expert and his supervisor during the drafting stage of the contribution show the differences of opinion between the two protagonists. According to the supervisor, the initial critique focused too heavily on the quality of the analysis conducted by the licensee, it was not sufficiently supported by risk scenarios that could occur at the facility, and these scenarios should have been identified during the review. This viewpoint is illustrated by the following comments the supervisor wrote about the identification of sensitive activities on several successive versions submitted by the expert: "It bothers me to say nothing about this identification. It sounds like there aren't any other activities" and "Are there any others that weren't identified by the licensee?" and "Again: are there any other activities?"

This repeated request would remain unheeded. From start to finish, the three sensitive activities addressed by the assessment were those identified by the licensee in its analysis. The supervisor's conception of the role of the assessment brings us back to the review stage. The expert considered that fulfilling his supervisor's request would be a costly endeavor requiring "several weeks of immersion in the licensee's facility", adding that it was not part of his job: "My job is to assess the licensee's analysis, not do the analysis for the licensee." (interview of March 9, 2006). Added to this is a comment made by a CEA human factors expert whom we spoke with after the assessment: "At the beginning, I had the feeling that the expert wanted to do something along the lines of a second opinion. He especially wanted to talk with the operating personnel. Such an approach is legitimate when the licensee provides nothing, but I'm not sure it's the right one in this case. We conducted an analysis of Minotaure. To me, the IRSN's role is to judge our approach. I think that the assessment made by the IRSN's human factors expert was quite on the mark." (interview of July 13, 2006). The human factors expert's supervisor's reply? "How can you perform an assessment without doing a confirmation analysis of the critical points?" (interview of September 5, 2006)

Lastly, one must not forget that the assessment covered the principles selected by the licensee to renovate its facility. It was for this reason in particular that the generalist engineers did not encourage the human factors expert to interview the operating personnel, which would have been a first step toward an immersion review. We will not delve any further into this attempt to find a common ground. Nevertheless, it is fundamental, recurrent and, it seems, not limited to the domain of human factors. "I must not perform the analysis for the licensee" is a maxim heard throughout the IRSN, where it is often repeated that it is the licensee's responsibility to demonstrate the safety of its facilities.

More interactions during the scope definition and review stages may have made the proofreading process less costly for both protagonists. Here we encounter a limit in our iconography, which may not have sufficiently explained this cost. Between the first version and the version sent to the coordinator, the expert wrote a total of six versions all while conducting other assessments. These six versions were proofread and commented by the expert's supervisor, who also had to proofread the assessments written by his ten employees in addition to performing his daily management duties. In all, the process dragged on for nearly four months. The proposed representation allows us to explain a few writing principles suggested by the supervisor to the expert. Of these principles, the supervisor was uncompromising about the following:

– Conclude each topic by providing an opinion consistent with the value (weakness, strength) and order of the argumentation. In other words, a generally positive critique must end with a statement along the lines of "this does not require any comment" and a generally negative critique must end with a request (cf. the topics of activity planning and interfaces).

– Requests must not be directly related to resources (cf. particularly in the case of the organization of feedback and the validation of design choices when the human factors expert would like the licensee to use human factors experts).

– The presentation of the licensee's case and the expert's opinion must be clearly separated.

A final less specific principle but which allows more room for negotiation in its implementation can be added: "Clarify the requests by showing the 'safety challenges' in their implementation". During the proofreading stage, the supervisor asked the following questions: What should one expect from an assessment of a new organization? What topics should be addressed by this assessment? What should the human factors chapter, which should be included in the annual safety report, address? Although the expert initially wanted the licensee to "set out its skills management considerations in an action plan", his supervisor had him clarify the topics of the action plan (training materials for new staff and number of tutors). These topics have to be related to safety. The fact that he did not see a clear link between the reorganization and safety explains why he was not convinced by the expert's request.

Thus, contrary to what one might have expected, the proofreading process thus does not boil down to merely validation. Regardless of the assessment or expert, the supervisor applies the aforementioned principles for it is vital that he familiarizes himself with the expert's contribution and be convinced of the requests made in order to be able to support them. Indeed, it is likely that the team of human factors experts has to face reluctant licensees. Internally, they also must convince the generalist engineers, who have to include the human factors contribution in their report.

4. The Transmission Stage (December 2005–March 2006)

In early December, the generalist engineers inserted the contribution in the final report and discussed the draft requests with the human factors experts. These draft requests were presented to the reactor safety director during an internal preparatory meeting. The draft of the final report was then sent to the licensee before being discussed during the preparatory meeting, which served as the final comparison before the meeting of the advisory committee. Initially scheduled for December 2005, it was postponed until March 2006. The most conflicting requests were discussed during it. Crucial steps, from the drafting of their contributions to the reporting of their conclusions to the representatives of the ASN, were performed by the assessment stakeholders.

4.1. Transmission of the Contribution to the Generalist Engineers

The generalist engineers embarked on an important and laborious stage in which they had to combine the thirteen expert contributions with their own analyses in one report. This was no mere task of piecing words together: "Our job is more than just pasting paragraphs together. It takes a lot of work to synthesize everything into a coherent whole." (interview of the unit head of April 11, 2006). This view was echoed by the coordinator, whose job it was to create this coherent whole from contributions that, on occasion, lacked in quality.

The human factors contribution was very well received by the generalist engineers. Said the unit head in an interview held on April 11, 2006: "I was very satisfied with this contribution. It surpassed my expectations, especially since it discussed reactor safety – such as with the double-check analysis – and not just ergonomic, psychological or training aspects. I've always said that the SEFH's experts should be skilled in reactor safety. We have taken a big step forward towards the human factors safety analysis." For the unit head, the added benefit this contribution had over the previous contributions was the analysis of sensitive activities in connection with reactor safety.

The handling of the requests was the subject of a meeting and a number of discussions between the team of human factors experts (the contributor and the supervisor) and the generalist engineers (the coordinator and the unit head). Two requests – that on the human factors chapter in the annual safety report and that on clearance management – were subsequently dropped. According to the coordinator "they were removed due to a standardization of the requests. The assessment addresses the approaches. The request on clearance management was too specific; we'll put it back in for the following assessment. We weren't really convinced by the first request. We discussed it with the human factors experts, and they agreed to take it out. Had they refused to do so, perhaps we would have left it in." (interview of February 22, 2006)

Before elaborating on the changes made to the human factors experts' other draft requests, we must add a fine distinction made during a preparatory meeting. The IRSN's draft requests were presented during this meeting. The licensee had two options: either to "consider the request as a commitment" and undertake to implement it or to reject it and see it turned into a draft recommendation. The draft recommendation would then be presented by the IRSN before the advisory committee. The licensee would be able to explain its refusal verbally and an open debate would follow. The committee would then decide whether to accept, modify or abandon the draft recommendation. The committee would then submit its recommendations to the ASN's representatives, which would decide whether to turn them into requests for the licensee.

The importance of this distinction became apparent during the preparatory meeting. However, thought was already being given to how to identify the various draft requests for organizational reasons. "No more than ten draft recommendations may be discussed during a session of the advisory committee", said an IRSN official during an interview held on May 12, 2006.

During their meeting with the human factors experts, the generalist engineers suggested grouping requests together. The discussions resulted in the establishment of five draft recommendations and two draft commitments (cf. Sidebar 11).

Draft recommendation R1

The advisory committee has duly noted that the facility's control room is undergoing a complete renovation and therefore recommends that the licensee provide the following information to the ASN:

– its method of drafting the documentation related to operation of the control room (procedures, operating instructions, etc.) in particular regarding the ergonomics validation stage with the future users,

– its method of identifying the skills required to work in the renovated control room and the training required to allow all staff (current and new) to acquire these skills,

– the validation results of tests of the new instrumentation and control arrangements of the facility's systems. The data collection and analysis program (tests and associated scenarios) shall be specified for each issue reviewed during testing.

Draft recommendation R2

The advisory committee recommends that, within six months' time, the licensee present to the ASN the measures it intends to implement to improve the consideration of human and organizational factors in the analysis of anomalies and incidents that occur in the facility.

Draft recommendation R3

The advisory committee has duly noted that the licensee has identified reactor operation and loading rod handling as "safety-sensitive activities". The advisory committee therefore recommends that, within six months' time, the licensee perform and provide the ASN with a human factors analysis for both activities. This analysis shall specify the potential human errors, their source factors and the measures to prevent them, detect them and mitigate their consequences.

Proposed recommendation R4

The advisory committee has duly noted that the licensee plans to implement physical and organizational measures to prevent the occurrence of the criticality risk in the loading room and, in particular, regarding the transfer of fissile materials from store MG1 to the loading room. The advisory committee has noted that the licensee undertakes to inform the ASN, within six months' time, of the concrete methods used to implement these measures and to explain their adequacy and robustness with regard to the criticality risks.

Proposed recommendation R5

The advisory committee recommends that the licensee take the necessary measures to improve the criticality training of all workers involved in the design and fabrication of loading rods.

Proposed commitment E1

The advisory committee has noted that, for the purposes of ascertaining the effectiveness of the newly created department, the licensee undertakes to, following complete unloading of the neutron monitoring system qualification core, present its conclusions on this organization and to specify and justify, as necessary, any actual or planned readjustments.

Proposed commitment E2

Regarding the renovation of the facility's control room, the Advisory committee notes that the licensee undertakes to draw up training materials suited to the needs of newcomers and the refresher training process. It also notes that the licensee undertakes to ensure that trainer turnover rates are compatible with the training needs of newcomers.

Sidebar 11: The draft human factors requests on December 13, 2005.

4.2. The Internal Preparatory Meeting

At the request of the IRSN's reactor safety director, the proposed commitments and recommendations were presented on December 16 during an internal preparatory meeting attended by all of the contributors and their supervisors. A total of twenty-one experts were in attendance and each contributor spoke in turn. The human factors expert spoke first, presenting his draft recommendations and commitments.

He started off with draft recommendation R4 (see above), to which the director replied: "Shouldn't the human factors aspects on criticality be pooled together with criticality during the advisory committee meeting? I prefer a presentation by risk type. Regarding criticality, there are sometimes organizational barriers and sometimes technical barriers. The advisory committee should see that the IRSN can work in a cross-disciplinary manner. I wouldn't want to see two separate chapters on the same subject." The human factors and criticality supervisors accepted the suggestion. The human factors and criticality experts would work together and write a joint chapter.

The human factors expert then presented recommendation R5 (see above), adding that the training lasts one-half day and that he didn't have any information about its content. The director then spoke of criticality incidents that had occurred at another facility and stressed the lack of criticality knowledge of the facility's new employees. He approved the recommendation, stressing the need for concrete, not "abstract or general", training. The human factors expert then moved on to the draft commitments. These were requests that the human factors experts and the generalist engineers had "wagered" would be readily accepted by the licensee. As there was no reaction from the audience, the drafts were maintained.

At the end of the presentation the director spoke of the radiation protection risks identified during a tour of the facility the week before: "I asked one operator how access to the reactor is restricted based on the operating conditions. His reply was not clear. EDF uses physical barriers, padlocks and chains. For me, these access issues are important. Perhaps the instructions are not sufficient. Perhaps technical barriers should be used in addition to organizational barriers." He then asked the experts if they had addressed these issues. Their answer was no, they had not. Disconcerted, the human factors expert replied that it was a "human factors issue" that he should have addressed. However, since the review was over, the issue would have to be addressed during the next review of the Minotaure reactor.

4.3. Transmission of the Draft Report to the Licensee

As suggested by the reactor safety director, the portion of the human factors expert's contribution on the risks related to fuel rod fabrication was included in the criticality experts' contribution. Draft recommendations R4 and R5 were inserted in the extensive draft recommendation R9.1 (cf. Sidebar 12).

> **R9.1** The advisory committee recommends that, within six months' time, the licensee present an analysis of the prevention of the criticality risk caused by human error. Such analysis shall detail the incident scenarios considered and the measures to be implemented to prevent such risk and demonstrate their adequacy and robustness in pursuance of FSR I.3.c (double contingency principle). For purposes of this analysis, it recommends that the licensee:
> – take into account a human-induced error occurring when fissile elements are placed in or removed from store MG1 (in particular, a zoning error during repositioning in the store, an error in the type of fissile element removed from the store, an error in the number of items removed from the store and taken to the fueling room, etc.),

– review the risk of errors caused by differences in the mass limits for fissile materials according to the criticality units in the fueling room (the mass limits for fissile materials in criticality units 12 and 18 differ from those in criticality units 5 to 11),

– given the various materials placed in store MG3, take into account an error occurring during the removal of materials therefrom (error in the counting of items, the types of materials placed in the fueling room, etc.),

– examine the criticality risks for the decontamination room and the hot laboratory and in particular specify the preventive measures for ensuring that the maximum limits authorized in these rooms for fissile and moderating materials are not exceeded,

– analyze the risks associated with failure to follow the instructions during transfers.

Furthermore, the advisory committee recommends that the licensee take the necessary measures to improve the criticality training of all workers involved in the design and fabrication of loading rods. Within this framework, the licensee shall specify in the GOR the necessity for training all new staff and the selected frequency of refresher training.

Sidebar 12: Draft recommendation R9.1 (January 25, 2006).

The generalist engineers sent the draft report to the licensee on January 18, 2006. The human factors experts and the generalist engineers made minor corrections to the draft requests up to the last minute.

4.4. The Comparison Meeting with the Licensee

On February 1, the human factors expert informed his supervisor of his apprehension about the following day's preparatory meeting with the licensee to discuss the draft commitments and recommendations. The department head therefore suggested a discussion during which the experts would prepare the comparison meeting. The human factors expert's apprehension was caused by the lack of a comparison meeting between the end of the review and the preparatory meeting. "This is the first time that I have not been able to discuss the draft chapter with the licensee. The coordinator was against it. Usually, the licensee has already seen our analysis. We don't spring it on the licensee during the preparatory meeting!"

As we have already seen with other assessments, the licensee and the expert sometimes come together before the preparatory meeting in order to discuss the contents of the report and clear up any misunderstandings. No such meetings were held in the case of Minotaure. "Relations were strained at the end of the review, particularly regarding criticality and fire prevention as well as the operating conditions," said the coordinator. "We preferred to end the review without holding these feedback meetings." (interview of February 22, 2006)

During their meeting, the human factors expert informed his supervisor of his apprehensions. The expert told him about the lack of discussions and that the licensees could mention this. The department head replied that it was the "usual misinformation" and that he should nevertheless inform the generalist engineers of it at the end of the advisory committee meeting for the review feedback. He also reviewed the list of draft requests and read portions of the arguments for each. For him, everything "held together". This is one outcome of the lengthy proofreading process. The supervisor, convinced of the suitability of the draft requests, would guide the human factors expert and stand by him in the event things became tense with the facility representatives during the following day's preparatory meeting.

On February 2, thirty people attended the meeting at the IRSN. In addition to the IRSN's experts, fifteen CEA representatives were present. The debates were heard by the ASN's representatives and a few members of the advisory committee.

Nearly half of the topics had been addressed the preceding day, when the meeting was first called to order. When the human factors experts arrived at 10 a.m., the containment risks were being debated. Upon seeing them arrive, the head of the criticality department discreetly approached them, saying "You're up first. That shouldn't be a problem because the CEA's criticality expert is well aware of human factors issues." Together, they briefly discussed the joint recommendation. At 10:45 the coordinator announced that the criticality risk was the next item on the agenda. The discussion thus turned to draft recommendation R9.1. The mood was tense. "Before starting I'd like to say that we received no news between June and January," said the licensee. "We were surprised to receive chapter 9 [on the criticality risk]! I'm telling you this so that you know what frame of mind we're in." No one spoke. "We're going to tell you what we think," said the licensee. "Our approach during the review was constructive. We analyzed the human errors. We identified and reviewed the measures for preventing these errors. We reinforced them regarding our operating staff in the loading room. We're surprised by the recommendation's relatively strong tenor. We worked with a human factors expert on this matter, so we're having a hard time understanding some of the scenarios identified. Some were kept while others were excluded because they were not deemed relevant. A number of barriers allow us to consider that they indeed are not. We therefore propose restating the recommendation as follows …. In short, we understand that there are things that need to be fleshed out and that training needs to be formally set out."

"That's exactly what we expect," replied the head of the criticality department. "Perhaps we'll read our proposed recommendation. The key terms are adequacy and robustness." To this, the licensee replied, "This is an advisory committee of principles, there are difficulties, and it must be said. We received no feedback! That isn't what technical dialogue is about! We could have discussed it together. If we're going to agree on a recommendation, we also have to agree on its underlying basis."

During this somewhat disjointed discussion, the licensee pointed out that "... we worked with the operators. This is a major effort. The operators are now aware and double-checks are being performed."

At this point the head of the SEFH attempted to redirect the discussion by commenting the draft recommendation, as seen in the following exchanges:

– The head of the SEFH: The work being done for the loading room is considerable compared with what is being done at other CEA facilities. That's one thing. There are two parts to the recommendation: 1) the scenarios, which concern criticality, and 2) prevention, which concerns human factors. I can understand that you find this rough. What is important is that the recommendation shows the progress that remains to be made. There are two aspects: 1) How is it that some scenarios have been excluded...

– The licensee: That's exactly what has not been discussed.

– The head of the SEFH: ... and 2) prevention principles have been put forward, but we need additional information. The term "flesh out" suits me fine. But, in your suggested recommendation, the wording "present the elements" is a tad inadequate. You need to talk about robustness.

– The licensee: We're in agreement on the first point. Robustness is obvious! Right!

– The head of the SEFH: Indeed, except that it's never done!

– The licensee: If we don't talk to each other, we'll never make any headway. We can't justify everything we haven't accepted.

At 11:15 the representative of the ASN spoke: "Right. The CEA will come to an agreement and then propose a commitment in line with the IRSN's comment?" The head of the SEFH then began dictating to a person taking notes on a laptop:

– The head of the SEFH: I propose that you 1) justify the non-sensitive character...

– The licensee: We'll do it for what is non-sensitive but not for what is virtual! Some scenarios are highly hypothetical!

By now it seemed as if the disagreement had been resolved, but then the head of the SEFH pronounced the following word:

– The head of the SEFH: 2) ... robustness ...

– The licensee: No, no, no! We won't accept that! Robustness isn't just something that is said; there are written instructions.

At this point, a participant suggested the term "relevance".

– The head of the criticality department: But that's not the same thing! Relevance means appropriateness. Something can be relevant but fragile!

Ultimately, he changed his mind, allowing the coordinator to close the discussions and move on to the following draft recommendation. In the end, draft recommendation R9.1 was replaced by commitment E9.1.

At 12:45, in the midst of the discussions, the facility manager took a break and left the room. The human factors experts followed and began discussing the requests:

– The head of the SEFH: What about our commitments?

– The licensee: OK for R11.1. OK for R11.2, but from the date of the safety analysis report [only the deadlines have been changed]. It has to be made consistent with R9.4 on handling activities. As for R11.3, we accept it for 18 months for the first two bullets. What was it for the third bullet?

– The head of the SEFH: They want to make you explain your method.

– The licensee: OK, but they'll have to give us time. As for E11.1, we're not good at organization operations evaluations. We'll give you some feedback, OK?

– The head of the SEFH: OK.

Upon their return to the room, the discussion moved to human factors. The head of the SEFH read all the recommendations accepted by the licensee modulo the agreed corrections, to which the head of the generalist engineers pointed out that the draft recommendations were now commitments.

The human factors expert left the preparatory meeting reassured. Instead of the disaster he was dreading, everything went along smoothly and the licensee committed to all the draft recommendations.

4.5. The End of the Assessment

Two daylong meetings of the advisory committee (March 9 and 16, 2006) were necessary to discuss twenty proposed recommendations that were not "translated into commitments". After the second session was adjourned, the advisory committee chairman sent his opinion to the director of the ASN. An initial first version of this opinion was proposed by the IRSN's generalist engineers (the section relating to human factors is provided in Sidebar 13). These conclusions were incorporated in a letter sent by the ASN to the licensee on June 14, 2006. The advisory committee's recommendations were translated into requests and the licensee's commitments were listed in an appendix.

"The advisory committee welcomes the improvement in the consideration of criticality risks in the facility, particularly in terms of scenarios reviewed and models adopted in the studies. That said, additional information must be provided in order to justify the criteria adopted and explore a wider spectrum of configurations in terms of the types of items, discussion conditions and geometry.

As regards prevention of risks related to human and organizational factors, the advisory committee notes that the licensee will provide additional information on human errors that may arise in the course of the various sensitive activities (fabrication of fuel rods in the loading room, fuel rod handling, reactor operation) and specify the measures taken to prevent such failures ....

The advisory committee is in favor of the measures adopted by the licensee to proceed with the safety review and renovation of the Minotaure facility provided the assessments and recommendations herein are taken into account ... and provided the licensee fulfills its commitments."

Sidebar 13: Excerpt of the opinion sent by the advisory group to the ASN. (April 19, 2006).

5. PROVISIONAL SUMMARY

5.1. The Basic Operations of the Human Factors Assessment

At the start of the assessment, the human factors expert works with the coordinator, who provides him with the necessary information about the facility (context, hazardous activities) and is his main go-between with the licensee. During the scope definition stage, a number of key meetings (milestones) held during the assessment can significantly impact the course of the human factors assessment (negotiations about documents provided by the licensee, licensee reply deadlines, procedures). It was during one such meeting that the licensee denied the human factors expert's request to interview the facility's operating personnel.

During the review stage, the human factors both consulted documents provided by the licensee (notably a human factors analysis) and obtained data from the licensee during two meetings and one facility tour. While following this stage, we made the following observations:

- *Some topics of the assessment emerged during the review (e.g. reorganization, activity planning, field interfaces). The entire range of areas assessed was therefore not set in stone at the start of the assessment. In fact, it depended especially on the expert's interactions with the licensee and the data provided;*

- *However, on a number of occasions the expert either stated or alluded to standard principles (e.g. "operating experience feedback must be set out in writing" and "the documentation must be validated"). Although such was neither clarified a priori nor shared with the other human factors experts, the expert had in mind a frame of reference, at least partial, that seemed to be independent of the facilities assessed.*

- *The expert attempted to evaluate the skills of the staff called on by the licensee to carry out a number of actions (e.g. drafting of the human factors analysis and operating documentation).*

- *In order to be able to assess the fuel rod fabrication activity, the expert asked to have the process and the work and checks performed by the operators explained to him. He then attempted to construct the chains of events that could lead to a hazard before evaluating the measures implemented. To do so, he was able to count on the support of the coordinator and the criticality expert.*

- *The licensee's representatives were cordial throughout the review. They also tried to persuade the expert that human factors had effectively been addressed, failures had been identified and human and organizational barriers were in place. But, as has been seen, a letter from the heads of the generalist engineering and criticality indicating a possible restriction in the facility's operation made it possible to obtain an analysis of the risks related to fuel rod fabrication.*

While writing his contribution, the expert received requests for revisions from his supervisor, who considered that the argumentations in the initial drafts did not sufficiently address safety at the Minotaure facility and found the review to be incomplete. In response, the expert fleshed out his review of a number of topics and polished his argumentations by mentioning past incidents and referring more to the licensee's case. He also withdrew some of his draft requests. At the end of this costly stage, the contribution of the SEFH and the draft requests were sent to the generalist engineers.

The contribution was easily incorporated by the generalist engineers into their draft report. The expert presented the draft human factors requests during an internal meeting. They were deemed legitimate by the reactor safety director, who wished for presentation efforts from the human factors and criticality experts and identified a topic – radiological protection – that had not been assessed. A few weeks later, during the preparatory meeting, the licensee accepted the draft requests after a few changes were made to the deadlines and terms used (robustness-relevance, evaluation-OEF in particular) and agreed to carry them out. The experts' role in the assessment ended here.

As stated in the general introduction, we would like to compare the human factors assessment on the review of safety at the Minotaure reactor against the three assessment models used and Ouchi's various forms of control.

5.2. Comparison of the Empirical Data and the Theoretical Models

The various basic operations set out render the canonical model ill-suited to characterizing the human factors assessment. Indeed, a number of its proposals are invalidated by our empirical data. In the first place, the human factors contribution analyzed is not an individual assessment. In fact, we were able to observe its many collective aspects (horizontal and vertical relationships between the experts, interactions with the licensee in particular). The second characteristic that further distances the human factors contribution from the canonical model relates to the knowledge used and created. First of all, the assessment process cannot be considered as separate from the expert's learning. For example, we saw that one part of the review was devoted to exploring event chains that might lead to a hazard (fuel rod fabrication analysis). This stage is a source of information for the expert because it allows him to establish connections between the human and organizational variables and safety that might be used for future cases. This is because the expert draws on his own experience during the assessment process. As the expert said when we asked him to explain his draft request on reorganization during an interview, "Reorganization can degrade safety at a facility. I've seen it happen in other cases." (September 13, 2006).

The case analyzed also shows that the expert did not have much freedom of choice of procedures, as he was not allowed to interview the facility's operating personnel. The occurrence of contingent operations (such as the negotiation of draft requests with the licensee) leads us to state that the assessment was dependent on the expert, which distances it a little bit more from the canonical model. Indeed, the outcome of the discussions depended notably on the expert's negotiating skills. It can be assumed that not every expert has exactly the same talent. The assessment's "path dependence" is also illustrated by the emergence of the topic of radiological protection, which had not been addressed by the human factors experts but mentioned by the reactor safety director during a meeting that initially had not been scheduled.

The many procedures in the assessment, and particularly those on its validation, can be seen as ways of reducing this dependence and illustrate the procedural aspect of the case studied. Yet, although these procedures integrate the adversarial principle, it is harder to associate them with the principles of independence and transparency.

But what happens when we apply Ouchi's theory? Do our data identify forms of result-based or procedure-based control or forms of clan control (control of levels of competence)?

Even though it is not clarified a priori, the expert's frame of reference is the best tool of bureaucratic control, which was used several times during the review stage. Drawing on this frame of reference effectively amounts to acknowledging that implementing barriers associated with the underlying principles (e.g. OEF set out in writing, validated documents) contributes to guaranteeing safety in the facility.

However, this was not the only form of control that was observed. The expert's attempt at evaluating the skills of the staff who carried out the human factors analysis, incident analyses and control room renovation project was more akin to a form of clan control. When he made reference to past incidents, his assessment resembled a form of result-based control.

Although some of the expert's interactions with the licensee can be identified with forms of control, they are not enough to characterize all of the operations in the review stage. This is because the analysis of the rod fabrication activity is not akin to a type of control but rather a real investigation that strove to reconstruct causal relationships that link together basic operative procedures.

Our observations show that the experts enjoyed full freedom in making their draft recommendations. That said, the compromise that followed the negotiations between the experts and the facility manager resulted in the assessment's conclusions being reformulated. Isn't it surprising that the licensee was able to have a say in the deadlines and even the content of requests it received? And what about a move by the licensee that limited the expert's freedom of action? Isn't it surprising that the licensee was able to influence the expert's work by refusing to allow him to privately interview the operators and by asking him to justify his methodological choices?

Chapter 4. The Analysis of Incidents at Artémis

"There is no science of the accident."
Aristotle

Like the assessment of the Minotaure reactor, the second case we studied concerned a research facility operated by the CEA and named Artémis (cf. Sidebar 14). However, in this case, the human factors experts' assessment was quite different in terms of the resources used and its conclusions. Following an in-depth investigation of two incidents that occurred in 2004 and 2005, the expert mapped out the chronology and the "causal relationship" of each. He expressed his opinion on the progress of the corrective actions the licensee had undertaken to implement and proposed others.

Between 1999 and 2004, six incidents occurred at Artémis. A seventh incident, which occurred in June 2005, was the subject of a letter from the ASN to the IRSN and dated October 17, 2005. The sending of this letter was accompanied by a strong appeal to the IRSN's experts by a representative of the ASN, with whom we spoke a few days afterward: "This series of incidents must be put to a stop. This time, we want an assessment and we want the human factor experts to take part in it." (interview of November 29, 2005)

Why did the ASN insist so much on the need for involving human factors experts? The ASN representative in charge of the case, and who had in his possession the incident reports submitted by the licensee and had participated in several inspections of the facility, considered that the incidents were caused by one factor in particular: the licensee/researcher interface. Here is how he explained it during our interview: "Artémis' operating staff belong to a department that makes the facility available to researchers from other departments. There is an interface problem between the licensees, who are responsible for safety, and the researchers. In 2004, the researchers went so far as to touch filters when in fact they should have informed the licensee. In 2005, they made pipe connections that should have involved the licensee. There is an impression that the researchers take many liberties. The 'lack of safety culture' explanation doesn't satisfy me. That's why I want an assessment. One shouldn't start with too much bias; I expect a certain degree of objectivity. Those are the rules accepted by the SEFH's experts to properly understand the interactions." (interview of November 29, 2005)

Thus was one of the challenges of the request made by the ASN's representatives: better understand the licensee/researcher interface with a guarantee of an unbiased analysis method. This lack of bias is guaranteed in particular by the attention given to the various stakeholders of the incident, regardless of their function. In order to properly understand the motivations of the ASN and IRSN's representatives, one must know that the direct contacts at Artémis belong to the facility's operating staff, who are in charge of its safety. They have very little contact with the researchers apart from brief questions they may ask them during inspections. Interviews of the facility's various 'populations' is thus an important source of information.

In addition to wanting to put an end to this series of incidents by better analyzing the human and organizational factors involved, an additional argument was put forward to justify the need for conducting a human factors assessment. Artémis' operating staff was in the midst of a safety review and would soon provide the ASN with an updated safety analysis report that would be assessed by the IRSN the following year. Conducting an assessment of the incidents put pressure on the licensee to adequately take human factors into account in the safety review.

Thus, after an assessment planning meeting, the generalist engineers in charge of the Artémis assessment asked the head of the SEFH to view the Artémis incidents as a priority.

At the SEFH, a human factors expert in charge of assessments at laboratories and plants was thus called on. His workload was reviewed with the agreement of the generalist engineers in charge of the facilities in question and it was decided that his contributions to safety reviews at two other facilities would wait. The expert was a young ergonomist who had been at the SEFH for three years.

As in the previous case, the human factors expert had everything to gain by working closely with the generalist engineer. Unlike the Minotaure safety review, for which a team of experts was formed, they were the only two experts involved. Nonetheless, this did not relax the formality of their collaboration. Whereas one might expect a duo whose mission would be to write a joint assessment – which could make the proofreading process easier – such was not the case. After receiving an official request, the expert had to submit a contribution to generalist engineer, who included it in a validated report that was sent to the ASN.

The licensee/researcher interface was not the only factor explored during the Artémis assessment. Various avenues to be pursued were identified during the scope definition stage.

> The Artémis facility is dedicated "to the R&D needs relating to the reprocessing processes, the processing and conditioning of waste, the chemistry of transuranium elements and the analysis methods associated with these methods." The facility has several laboratories where experiments are conducted on radioactive materials using gloveboxes or, in the case of solutions of higher radioactivity (beta and gamma radiation), remote handling equipment. The solutions are then confined in shielded housings isolated from the exterior by two thick layers of lead and steel. The housings are assembled together to form a shielded line. It is a large facility with a large workforce (around 250) of various groups of people (researchers, PhD students, laboratory assistants, operating engineers).

Sidebar 14: What is Artémis?

1. THE SCOPE DEFINITION STAGE (OCTOBER–NOVEMBER 2005)

During the scope definition stage, the experts discussed amongst each other the areas they felt needed to be assessed. The scope of the assessment and the weight to be given to the five incidents that had occurred prior to the 2004 incident were discussed by the experts and the generalist engineers. The extent of the interaction between the expert and his supervisor prior to the assessment and the highly contextualized nature of the areas identified contrast with the previous case.

In late October 2005, the human factors expert had just finished an assessment and was ready to devote himself entirely to assessing the incidents that had occurred at Artémis. The impressions that he shared with us during an interview reflect the existence of what could be termed as the "initial cost" of the assessment: "I accepted the case. I have a tough time moving from one case to another. Three years of experience is not a lot. I'm not at all familiar with Artémis, its organization or what is done there. I have to start from zero, which requires a lot of work. I tend to want to properly understand things before talking about them. For example, take the incident that I just finished analyzing. I can say what happened and I can describe the facility. But here, I've got to start from zero." (interview of October 27, 2005)

He had in his possession the significant incident report that the licensee had submitted to the ASN four months after the 2004 incident. This document, some ten pages in length, described the sequence of the incident, the corrective actions that were immediately implemented, the findings of investigations conducted to determine the causes of the incident, its real and potential consequences, its causes, and corrective actions identified through operating experience feedback.

As for the 2005 incident, the expert had in his possession the significant incident notification sent by the licensee to the ASN the day after the incident. This two-page document provided a brief description of the incident, its causes and consequences, the actions immediately taken by the licensee, and a proposal for ranking the incident on the INES scale. A few days later, the coordinator provided the expert with the significant incident report for the 2005 incident. The report's structure is similar to that for the 2004 incident.

He also had the report of a "reactive inspection" that had been organized by the ASN's regional division and in which the IRNS's coordinating generalist engineer had participated. The three-page report included a summary of the inspection and requests for corrective actions.

During his first reading of these documents, the expert noticed that the reports contained a number of deficiencies. "Incident reports are often very technical. This report is missing some information about human factors. For example, it doesn't say why the researcher didn't do what he was supposed to. Did he not have the necessary resources? What was he doing? If it contained more information about human factors, we would have more specific questions and we could start the analysis much more quickly. For example, it says 'an operator had prepared the check ... but forgot to provide it to the other operator.' One can wonder why he forgot to do so. If I read that a check was not performed, I'm going to ask why. 'Because he forgot' is the typical answer. I'm going to have to ask why again ...." (interview of October 27, 2005)

The expert planned to prepare a questionnaire that would allow him to analyze both incidents. He also wanted to speak with his supervisor because he didn't know how far he could go in the analysis, wondering "Should I include the five previous incidents that occurred between 1999 and 2002 in my assessment?" (interview of October 27, 2005) He asked his supervisor that question the very same day. He would have liked to refer to the ASN's request, but it hadn't yet been received. At the end of their discussion, his question remained unsettled. An appointment was made to discuss the two incident reports. This discussion is briefly summarized below (Sidebars 15 and 16).

In March 2004, it was decided to denitrate radioactive effluents in order to lower their acidity before transferring to another facility for treatment. The process involves heating the effluents to be denitrated in an evaporator then successively injecting sodium nitrate and formic acid into the evaporator to destroy the nitric acid. Several operators performed this operation on the day of the incident.

Soon after formic acid was added to the evaporator, a responder from the Radiological Protection Unit (RPU), located in another part of the facility, noticed an increase in the amount of gas exiting the facility's stack.

Seeing that the denitration process was not starting inside the evaporator, the researchers lowered the flow of formic acid. Despite their action, in the minutes that followed the reaction ran out of control, automatically stopping the process and generating an alarm indicating the fouling of two filters on the extraction system.

Suspecting that liquid had reached the filters, they went to investigate. Seeing that the filter housings were covered in liquid, they contacted the RPU's responders, who came to investigate. The decision was then made to replace the filters. Equipment was moved so that a vinyl airlock could be erected around them.

Once the structure was erected, one of the researchers checked for contamination. He used a monitor from another part of the facility because the one at the exit of the room containing the filters was unavailable. The monitor gave a positive readout and all of the researchers were treated by the RPU's responders.

The five researchers became contaminated when they moved the equipment[167]. The pressure from the uncontrolled denitration reaction forced the effluents up the system and into the room via a leaky valve located upstream of the filters.

Sidebar 15: Summary of the 2004 incident (sources: CEA document).

In June 2005, a researcher used nitric acid to clean the worktop of a cabinet in which radioactive solutions were confined. According to practice, after rinsing the worktop, the used, and thus radioactive, rinsing solution had to be transferred to a tank inside another cabinet.

On the Friday before the incident, the worktop had been rinsed off. However, the rinsing line (consisting of connections between the two cabinets and a workstation with a glovebox) was left as-is until the following Tuesday instead of being disassembled immediately after rinsing.

On Monday, the used rinsing solution, left at the bottom of the rinsed cabinet over the weekend, was added to a container in order to be transferred to the tank of another cabinet the following day.

---

[167] According to the report, only two of the five researchers suffered serious internal contamination and "further medical monitoring revealed that the respective absorbed doses were ten times lower than the regulatory limits."

On Tuesday, the researchers assembled a line to perform the transfer. Upon seeing that the level in the tank was not rising, they stopped the transfer but, a few seconds later, the building's criticality alarm sounded. What happened is that the researchers had incorrectly assembled the line, causing the contents of the container to be transferred to the glovebox, which was still connected to the cabinet housing the container because the rinsing line used the previous Friday was still in place. However, the workstation was not equipped to receive solutions with such a level of radioactivity.[168].

Sidebar 16: Summary of the 2005 incident (sources: CEA document).

On November 4, the expert and his supervisor spent nearly one hour discussing both incidents in turn. The expert listed the areas that he wanted to review and his supervisor clarified or made changes to them as necessary.

The first area identified by the expert was human-machine interfaces. "In the case of the 2004 incident, I wouldn't frame the issue in terms of HMIs but rather overall equipment design," said the supervisor. "In particular, there's the issue of the overly long pipe[169] that seems to have been what caused the reaction to run out of control."

The expert proposed a second area by asking "What information do the operators have in order to monitor the reaction?" The supervisor clarified his question: "Once the reaction runs out of control everything apparently happens very quickly. You need to distinguish the and experiment set-up and visual inspection stages from the moment when the responders identify a problem with the filters. There are two things: the set-up and supervision of an experiment, and the rules to be followed under degraded conditions."

The expert completely agreed with his supervisor, who continued to lead the discussion and give him advice on how to prepare his questionnaire. "You need to ask questions about what the RPU's responders exactly did when they indentified the problem of the filters. Also, what is the procedure for managing abnormal situations? Is denitration carried out frequently? However, you're going to have to see the generalist engineer about the issue of design because that's beyond our expertise."

The discussion then turned to the 2005 incident. "After an experiment, they decided to clean a worktop. Apparently, they connected a pipe in the wrong

---

[168] The incident report goes on to state that there were no cases of immediate irradiation of workers and that no workers were near the glovebox at the time of the incident. It would be later learned that, despite cleanup measures, the incident resulted in a dose equivalent rate near the glovebox higher than the limit value set for a non-permanent workstation in the facility. An exclusion zone set up around the glovebox was still in place when the experts toured the facility on January 18, 2006.

[169] One of the causes identified by the licensee in its report.

place," said the supervisor. "... Right, and because there wasn't a check valve in place, the radioactive solution went into a glovebox. It's not easy to understand because the drawings aren't very clear," replied the expert. "Here, too, you need to ask questions about the set-up," said the supervisor. "And you need to understand who does what, who checks what, how the operators are trained and what documents they use. All this isn't clear in the incident report. Also, you need to know whether it's a modification that requires a procedure or an action that is part of normal operation. That would give us the tools to dig deeper into the issue of role-sharing between the researchers and safety engineering. I emphasize this aspect because the matter of safety engineering's role comes up repeatedly in the five previous incidents."

The supervisor recommended that the expert discuss these various points with the coordinating generalist engineer, who well understood the technical processes and organizational measures in force in the facility.

At the end of their discussion, they decided to call the generalist engineer in order to agree on the assessment's scope. However, the incidents were the first thing they discussed over phone. Regarding the 2004 incident, the generalist engineer recommended dropping the design aspect, saying that he would handle that himself. As for the 2005 incident, he stated that it wasn't a modification. "Nor is it a routine operation, but it should become one. At the time, they were in a start-up phase with a view to routine operation .... The set-up that led to the incident was not the original one; the pump was added by the researcher without any oversight by the facility manager." (generalist engineer, conference call of November 4, 2005)

The coordinator had a rather clear opinion, based on his experience, on the causes of these incidents. "To me, it's always the same thing. The researchers take initiatives to work around organizational, procedural or design shortcomings, and that leads to deviations beyond operational conditions." (conference call of November 4, 2005)

After this call, the expert told me that he had misgivings about the coordinator's viewpoint. To him, it was too cut and dried, and he didn't know what it was based on. He preferred to largely ignore it for the rest of the assessment. The experts finally got around to discussing the assessment's scope.

The generalist engineer considered that the experts should include the seven incidents in their contribution. However, finishing on time was the overriding objective. He was therefore willing to allow the experts a certain degree of freedom and mentioned the next "meeting": "In any event, the overall human factors assessment will be conducted during the safety review. The aim is to make a number of human factors requests so that the licensee is ready for reassessment. So, I'm going to make a request for what you can do." (the coordinator, conference call of November 4, 2005)

It was at this time that the head of the SEFH decided to call the generalist engineer's supervisor (unit head). After a short, cordial discussion, the experts opted together for an in-depth analysis of the last two incidents, followed by conclusions possibly containing references to past incidents. The human factors expert was happy with this decision.

Although the Minotaure review and the Artémis incident analysis differed in many ways, they did share a few of the same themes (activity planning, operator training, field interfaces). In the present case, these themes emerged during the work sessions between the expert and his supervisor (such sessions had not been held during the same stage of the Minotaure assessment). Both parties viewed these sessions as constructive. The expert benefited from the advice of his more experienced supervisor who, in any case, would validate his work. The supervisor expected a number of beneficial effects from these meetings. In particular, he hoped to quickly converge his views with those of the expert during the proofreading stage.

At this stage of the analysis, the following three themes had been identified:

- *Facility design issues (overly long pipe in the case of the 2004 incident; lack of a check valve in the case of the 2005 incident)*

- *Organizational measures regarding the set-up, supervision and monitoring of experiments (and the role of safety engineering and researcher training in particular)*

- *Management of degraded situations (intervention by the RPU's experts)*

2. THE REVIEW STAGE (NOVEMBER 2005–JANUARY 2006)

The two questionnaires that the human factors expert sent to the coordinator fueled the discussions between the two parties prior to their tour of the facility. After one such discussion, the expert told us of one his concerns: "I'm not going to be able to analyze both incidents in just one day. I'll need two days, but I'm a little afraid the coordinator won't agree[170]. On the other hand, I would understand if the licensee said no. The incident analysis is a hassle for the licensee and it has nothing to expect from it. What's more, the licensee may end up having to implement corrective actions! The priority is to show that our work can help them to better understand what happened." (interview of

---

[170] As in the previous case, the two experts had never worked together before.

November 15, 2005) Ultimately, the coordinator and the licensee agreed to allow the assessment to take place over two days.

Reconstructing the review stage of the incident analysis requires detailed explanations of complicated technical operations. For the sake of clarity, but at the expense of strictly adhering to the chronological order of events, we will explain the investigations of each incident first, followed by those of the corrective actions.

2.1. The 2004 Incident

The first items in the questionnaire that the human factors experts sent to the coordinator concerned organization at Artémis. The thirty questions about the 2004 incident were grouped under the following four headings, which corresponded to the incident's four episodes:

- *Uncontrolled formic denitration reaction;*

- *Monitoring;*

- *Visual inspection of the filters;*

- *Removal of the filters – Erection of a vinyl airlock.*

Portions of the incident report relating to each heading are contained in sidebars and followed by the experts' comments.

On November 15, 2005, the experts and the coordinator held a conference call. It began with a presentation by the coordinator of the organization at Artémis. "Artémis is overseen by a department that consists of the operations unit, run by the facility manager, and two units of researchers who work at the facility. There are also researchers from other units. Safety engineering belongs to the operations unit." The expert asked the coordinator to describe the nature of the relationships between the operations unit and the researchers. The coordinator replied by referring to the symbolic licensee/researcher relationships: "Each has a unit head. The head of the operations unit is in charge of the facility's operation and safety. The heads of the researcher units are in charge of the programs. So, yes, the department head's role should be to do both."

The expert asked for more information about the specifics of the denitration operation: "Is it a routine operation? How are tasks divided between the two researchers? What interfaces do they use to control and check actions? How was the operation prepared?" In addition to answering the expert's questions, the coordinator gave his opinion on what happened and what should have happened. "This was the first time since 1992 that they had performed this operation! Before that, they had no need to. This was the first time, that's why things went wrong. There was no preparation and there was no procedure despite the fact that they had never performed the operation before and that it's known to be tricky. I'm familiar with it. I know it's not something that you should improvise! What's more, they ran the operation in automatic mode when it should have been done manually. In automatic mode, the machine can't run fast enough. In manual mode, you have to monitor the pressure and cut off the heat as soon as things start to speed up. Usually, the system is equipped with a froth breaker [a cooling ring] to stop the reaction, but there wasn't one in this case. It's common knowledge that the pressure rises during this reaction." (conference call of November 15, 2005)

The conversation then turned to monitoring and particularly the interactions between the radiological protection experts and the researchers.

> Soon after the denitration operation, the responders in the RPU, located in another part of the facility, saw on the screen of the radiological protection panel an increase in the amount of gaseous discharges detected by the measurement sensors on one of the facility's stacks.

"Did they have any hypotheses about the causes of the discharges at that time? How did they look for the causes? Did the researchers receive any special instructions from the RPU?" asked the expert. "I don't think the RPU was aware of what was going on," replied the coordinator. "It just saw a rise in the measurements but didn't know exactly why. That's a question you'll have to ask. You'll also have to ask when the facility manager was informed."

The expert then asked a series of questions about the visual inspection of the filters.

> After the denitration reaction ran out of control, the researchers were alerted by an alarm indicating the fouling of two filters. Suspecting that radioactive effluents had migrated into the filters, they made the decision to visually inspect the filters.

The expert: "Who were these researchers? When did they decided to visually inspect the filters? Were they informed about the discharge identified by the RPU? How were the filters removed for the visual inspection? Was this task typically performed by the researchers? Did they follow a procedure? Was there a procedure? Should they have informed someone in particular before removing and replacing the filters?" The coordinator: "They were the same two researchers who had initiated the denitration reaction. In my opinion, there is no connection between the fouling alarm and the discharge identified by the RPU, but leave the question in."

The experts then moved on to the fourth theme – removal of the filters/erection of a vinyl airlock – which again involved the RPU.

> Seeing liquid on the filter housings, the researchers informed the RPU, which made two decisions: replace the filters and erect a vinyl airlock around the filters in order to do so. The RPU's responders first monitored for radiation around the filters since the zone had been identified as potentially contaminated. At this stage, none of the checks on the researchers revealed any contamination.

The expert: "Was the RPU's decision validated and tracked by safety engineering? Why didn't the RPU's responders have any surface contamination inspection equipment on them? Did anyone leave the area between the time the work area was set up (9:45 a.m.) and the time that it was discovered that a researcher had been contaminated (11:30 a.m.)? Did they check each other? Why did the operator check himself at 11:30? How was the erection of the vinyl airlock prepared?" The coordinator: "I'll rephrase one of your questions: how were the entry and exit of persons controlled between 9:45 and 11:30?" The expert: "Why did the operator have to remove his gloves in order to cut the fiber-reinforced adhesive tape?" The coordinator: "I can see you've never seen the tape that's used! Just try cutting it with gloves on! Still, you're right to ask that question. You'll have to ask whether the researcher got the OK from the RPU to take his gloves off."

At this point, the experts moved away from the questionnaire and discussed the event's aftermath and radiological consequences:

> Since no contamination had been detected, the RPU's responders left the room. The researchers began erecting the vinyl airlock and moved a number of objects. It was at this point that the researchers became contaminated. The pressure from the uncontrolled denitration reaction had forced the effluents into the ventilation duct and up to the filters. There, they spread into the room via a leaky valve located upstream of the filters. After the airlock was erected, a researcher checked himself with a monitor taken from another part of the facility. He discovered that his hands were contaminated. His four coworkers who participated in dismantling the airlock were treated by the RPU.

"They should have practiced with no-load denitration [with non-radioactive products]. A leaky valve is the type of thing you only see when something goes wrong. There are design flaws, particularly in relation to the location of the filters. And there's also the issue of the length of the formic acid feed pipe. But I'll take care of the analysis; you don't have to do it." (the coordinator)

The meeting wrapped up with a short discussion on the relationships between the licensee and the IRSN. The subject was broached when the coordinator responded to the expert's last question: "Yes, you'll have to ask if it's possible to have a copy of the instructions. I've never been able to get them in six years! They know the system inside out!" "Really?" asked the expert.

"How are relations with the licensee?" "They're good," replied the expert. "But you'll have to be very specific in your questions, otherwise they'll divert them ...."

At the end of the meeting, the expert was satisfied. The information provided by the coordinator had cleared many of the expert's misunderstandings. However, he admitted his fears about the relatively technical nature of the event being analyzed: "It's highly technical. I hope that they [the licensee] don't drive us crazy!" He also preferred to maintain a skeptical attitude about the coordinator's comments. "I tend not to take everything that generalist engineers say at face value. Some of the things [the coordinator] said were based on his impressions and feelings and can't be used as such. What's more, they always say their facility is different from the rest!"

The first meeting with the licensee, devoted to the 2004 incident, was scheduled for December 9, 2005. One week before the meeting, the expert proposed an agenda to the facility manager that included "interviews of the operators that had performed the formic acid denitration operation on March 11, 2004, (and who may be interviewed together) and the RPU's responders who had searched for the cause of the discharges and assisted the operators in erecting the vinyl airlock in order to replace the filters."

The agenda had actually been discussed beforehand with the coordinator, who subsequently proposed it to the safety engineer, his direct contact at Artémis. After consulting with his immediate superior, the safety engineer informed the coordinator of his agreement on the agenda. The expert saw this as the result of the coordinator's good negotiating skills: "The coordinator has a very strategic role. If I had written 'I want to meet researchers' on the fax, I don't think I would have gotten far. He went to them, explained my reasons, and they agreed." (interview of December 8, 2005)

The agenda was not followed in full. The first part of the morning (9:30 – 11:00 a.m.) consisted of a tour of the units in the facility where the incident occurred. The tour was attended by researchers, a RPU responder, the facility manager and the safety engineer. The expert did not consider it necessary to speak further with the incident's protagonists. At the facility manager's suggestion, the afternoon was devoted to a presentation of the corrective actions implemented.

The first part of the tour took place in the room where the two researchers had controlled the nitric acid feed. The facility manager showed the expert the mimic display used to supervise the reaction and pointed out the various interfaces used by the three[171] researchers. The researchers answered the expert's questions about how they set up the reaction. "The formic acid denitration operation was performed for the first time under hot conditions in the unit although cold tests had been conducted in 1992," said one researcher. "Documentation on setting up and controlling the reaction had been issued at the time and the test results had been recorded in an operations logbook. Also, the reaction had routinely been carried out around once or twice a year in another unit of the facility without any known incident. The researchers who performed the operation on March 11 were the same ones who had performed it in the other unit."

Later on in the conference room, the safety engineer provided the following clarifications: "For the facility's safety team, the formic acid denitration operation was not considered at risk for two reasons: the fact that it was relatively routine in the other unit, even though the risk of it going out of control was known, and that it did not deviate from the safety reference framework. It's true that, before performing this operation with a system that had never run under hot conditions before, the associated potential risks should have been considered even though the twenty-odd cold tests conducted in 1992 had been conclusive. At that time, the focus was on the risks related to the desulfation operations planned in the facility."

Whenever the researchers' descriptions became highly technical, the expert would consult the coordinator or ask the researchers to repeat themselves. "I'm just trying to understand," he would say to reassure them that they weren't being judged. Things became a little tense when it came time to discussing the method of heating. The coordinator considered that it was one of the main reasons why the reaction ran out of control ("if they at least had done it manually..."). There was no reaction from the facility manager or the safety engineer.

The expert then turned to the RPU responder for answers to his questions on the monitoring phase, which started when an alarm sounded. "At 8:50, we noticed the warning threshold alarm," said the responder. "We started calling the researchers, but not those in this unit. At 9:27, the alarm corresponding to threshold 2 began sounding. At that time, we didn't yet know what the source of the discharge was. Usually we're informed about operations that might cause releases. But the formic acid denitration operation wasn't considered as one, and so no one told us about it. At that moment, the researchers called us to

---

[171] Unlike what the experts had thought before the tour, three researchers, not two, had actually been present. One had been at the control station and the other two had been in charge of adjusting the various parameters.

inform us that the filters were fouled. That's why we didn't consider it necessary to inform the facility manager. Otherwise, we would have."

The group then moved to the room containing the filters. The expert asked the researchers what they did after the reaction went out of control. "We saw the filter alarms on the panels. We put on our face masks[172] and went to the transmitter room [containing the filters] to see what was going on. We saw that the filters were wet and called the RPU, but we didn't know about the release." (a researcher)

The expert asked if there were filter inspection instructions and if safety engineering had been informed about this operation. "There isn't a procedure for removing the housings. We simply put on face masks and gloves," said one researcher. "No, no special measures are needed to open and close the filter housings during visual inspections, even during an incident," said the safety engineer. "However, in order to remove the filters, the researchers had to contact the RPU, which is what they did."

The coordinator then asked if the researchers had checked themselves at that time. "We checked around for radiation before erecting the airlock. Because the body readings were negative, we focused on the filters but found nothing. We also checked for surface contamination around the filters (where the risk of contamination was highest), and again found nothing. When we finished our checks, we considered that the situation was not degraded and left. A hand-foot monitor in another unit upstairs was used to check for radiation each time someone left or entered the room containing the filters. We had to get another monitor because the one at the room's exit was out of order. It had become saturated by the uncontrolled denitration reaction that pushed in the effluents. We checked each other at least once and didn't find any contamination. The source of the contamination wasn't where it was supposed to be. There was no reason to monitor around the objects that had been moved by the researchers." (the RPU responder)

The expert turned to the facility manager, asking him how the intervention was coordinated. "A few days after the incident, we held a big meeting on its causes and on management of the post-incident phase," said the facility manager. "You have to understand that the situation was completely different from what we had expected."

Perhaps not entirely satisfied by this answer, the expert asked the RPU responder why he hadn't left the surface contamination monitoring equipment with the researchers while they continued their work. For the expert, this

---

[172] Face masks are used as protection against radiation.

equipment would have allowed the researchers to check themselves directly at the exit of the room containing the filters without having to go to the hand-foot monitors upstairs. "There was no need because no one suspected a risk of contamination in the room," said the RPU responder in an obvious tone.

The tour ended a little before 11:00 a.m. We then went to a conference room where additional answers were provided, particularly regarding the general operation of the Artémis facility. The licensee listed the various coordination meetings between the operators and the researchers. He dwelled on the position of operations manager, which he considered crucial. "Artémis is divided into forty-four operational units, each of which is overseen by an operations manager. The operations manager is the main channel of dialogue between operators, researchers and maintenance or building contractors. He is the main person responsible for locally enforcing the rules and safety reference framework that apply in a functional unit.... The operations to be carried out by the researchers are defined by the operations manager or the laboratory head[173] and are recorded in the operations logbook. The researchers define the resources to be used, with the agreement of the operations manager and/or the laboratory head during shift start-up meetings." (the facility manager)

For the coordinator, this mode of operation was suited to the facility and he told the expert as much: "This mode of operation is related to the type of activities conducted in research facilities, where a certain flexibility in the organization of experiments is needed in order for the facility to operate. Given the multiplicity of operations conducted there and the multitude of possible combinations in terms of connections and so on, it is necessary to leave the researchers a certain degree of latitude in performing their routine tasks in order to benefit from all the possibilities afforded by the equipment."

At the end of the day, the experts exchanged a few of their impressions at the facility's exit. "The organization at Artémis is even more complex that in a plant!", said the expert. To which the coordinator replied, "You heard the closed answers they gave. It's not easy." The expert agreed, but was nevertheless satisfied with the day's results.

2.2. The 2005 Incident

The 2005 incident differed from the 2004 incident in many respects. It occurred in a unit that contained one of the facility's shielded process lines. Researchers used manipulator arms to handle radioactive material confined inside shielded cabinets. The solutions were transferred from one cabinet to

---

[173] Head of a functional unit.

another through a system of tunnels. The researchers used special techniques to transfer the solutions to other units in the facility (vacuum transfers, transfers via packaging). To perform these transfers, they had to go to the "rear" of the shielded process line. A face mask was required during transfers of active solutions. The unit was also equipped with gloveboxes located above the shielded process line. The gloves in which the researchers placed their hands served as a containment barrier to protect them from the solutions being manipulated. To transfer solutions among the unit's various workstations, the researchers assembled a lineup using special pipes and couplings inside contained areas. The 2005 incident was caused by a pipe lineup error.

After conferring with the coordinator for a first time using his questionnaire, the expert reconstructed the 2005 incident for us during an interview. The incident's main "agents"[174] are as follows:

- *the workstations for cabinets C1 and C2 on the unit's shielded process line;*

- *the workstation of one of the gloveboxes (GBX) in the reagents room*

- *the three couplings used to transfer solutions between cabinets C1 and C2 and the glovebox.*

The human factors expert described the incident as follows:

"The week before the incident, a researcher cleaned the worktop of cabinet C2 with nitric acid. After the worktop was rinsed, the used – and thus radioactive – rinsing solution was supposed to be transferred to a tank inside housing C1. The worktop was rinsed off on Friday, June 24, using the following lineup:

– Pump 1 was installed in the glovebox containing nitric acid in order to feed the nitric acid to cabinet C1 via coupling r1;

– The acid was then transferred from cabinet C1 to cabinet C2 through coupling r2. The worktop on cabinet C2 was then rinsed off.

Instead of being disassembled immediately thereafter, the lineup was left in place until Tuesday, June 28.

On Monday, June 27, the used rinsing solution, left at the bottom of cabinet C2 over the weekend, was placed in a container in order to be transferred to the tank inside cabinet C1 the following day. The transfer was to occur using the following lineup: pump 2 installed in cabinet C2 in order to transfer the contents

---

[174] Given the significant role of equipment in this incident, this account uses this notion, which is used particularly by the sociologists of the Center for the Sociology of Innovation (C.S.I.) (see. Akrich, M., M. Callon, et al. (2006). *Sociologie de la traduction: textes fondateurs.* Paris, Les presses de l'Ecole des mines.) Furthermore, Nicolas Dodier (1994; 1995) clearly demonstrates how the concepts of sociology of innovation can be used to interpret occupational accident analysis methods.

of the container in C2 was supposed to be connected to coupling r3, which was connected to the tank in cabinet C1.

However, the following day, instead of connecting pump 2 to coupling r3, the researchers inadvertently connected it to coupling r2 between cabinets C1 and C2. Because the nitric acid lineup between the glovebox and cabinets C1 and C2 was still in place, it created a connection between the container in cabinet C2, which contained the radioactive rinsing solution, and the glovebox, which was not intended to receive any radioactive solution.

After turning on pump 2 and seeing that the level of the tank in cabinet C1 was not rising, the researchers halted the experiment. However, it was already too late. The building's criticality alarm sounded in the seconds that followed."

The diagram drawn by the expert to illustrate his account of the events is particularly enlightening in understanding the incident:

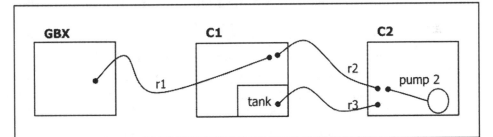

Sidebar 17: Diagram showing the connections between the glovebox (GBX) and cabinets C1 and C2.

On the second day of the review at the facility (January 18), the expert was again accompanied by the coordinator. Representing the licensee were, naturally, the facility manager and the safety engineer. There were also three researchers, one of whom was in charge of the operation of the functional unit containing the shielded process line.

The researchers provided information on how the task was organized. "All of the operations took place during six rotating shifts. The operators who assembled the lineup to transfer the nitric acid from the glovebox to C2 were not the same ones who assembled the lineup from C2 to C1. Three researchers were involved in creating the drain lineup: two researchers opposite the bottom of C2 were in charge of making the connection to C1 and one researcher opposite C1 was in charge of making the connection from C2. All three communicated with each other using walkie-talkies."

The discussion dwelled on the lineup error. The researcher simulated the error that he had made on June 28. Using the manipulator arm, he grasped the end of the hose of pump 2 and then showed us first the position of the coupling in the correct lineup and then that of the coupling that caused the error. The couplings were identified with labels. Turning to me, the expert said, "It's really hard to tell the labels apart. Also, you have to move to the very left of the window whereas you have to use the manipulator arm on the right to connect the hoses."

As the simulation continued, the researchers stated that "the ribbon cable at the bottom of C2 is asymmetrical with the ribbon cable at the bottom of C1 when in fact both are used to carry solution between C2 and C1." The expert noted that this asymmetry was not depicted in a drawing or mentioned in a procedure that could have warned the researchers about the precautions to be taken when making the connections.

Always in relation to what can be described as characterizing the researchers' working conditions, the safety engineer explained that the difficulty in telling the labels apart in cabinet C2 was because "the cabinet (measuring more than five meters high) is lit from the side by fluorescent lamps. The other cabinets are lit from the top, which casts more direct light."

In addition, the facility manager said that after the incident he identified that the configuration at the time made it possible to connect lines used for potentially radioactive products with lines used for non-radioactive products. Without check valves between the cabinets and the gloveboxes, it was possible for radioactive products to flow into the gloveboxes.

The expert and the coordinator then asked about the lineup checks, to which the facility manager responded by saying "The lineups weren't checked because these operations didn't deviate from the safety reference framework and thus were not identified as being hazardous."

As for the disassembly of the lineup used to clean the worktop, the researchers said "Before assembling the rinsing solution drain lineup, we didn't check whether the nitric acid feed lineup assembled on Friday was still in place. We thought that it had been removed…. It should have been disassembled after being used, but the researchers had left it in order to rinse the cabinet a second time."

The expert asked the researchers if there was a document on assembling lineups. "There weren't any documents on transferring solutions from the gloveboxes to the cabinets or on transferring solutions from one cabinet to another. That said, the lineup was simple. We can't guarantee one hundred percent that we would have referred to a procedure had there been one."

The facility manager and the human factors expert shared their impressions. The expert: "When you get down to it, it's a design problem. Had there been a check valve…." The facility manager: "It seems to me that it's more of an organizational problem. A problem of set-up and checking. It wasn't in the researchers' culture to check every lineup". In short, when the human factors expert focused on the technical aspects, the technical expert focused on human factors.

At the end of the day the expert once again satisfied with the working sessions. He wasn't convinced by the facility manager's presentation of what initiated the risks analyses. He considered that the operations should have been identified as hazardous. "They involved a breach of the first containment barrier. What's more, the pumps in the glovebox and at the bottom of cabinet C2 were a deviation from the reference framework. A modification request and a risk analysis would have been necessary before they were used."

2.3. The Corrective Actions

The incidents, and particularly the one that occurred on a shielded process line that had been in service only since 2002, provoked a reaction from the upper echelon of the CEA. Therefore, during the second half of a meeting with the IRSN's experts on December 9, 2005, the facility manager presented an ambitious action. The licensee proposed several types of corrective actions that consisted of an addition or modification of a technical system of the equipment implicated in the incidents as well as documentary modifications. But, because a strong human factor was identified following incidents, the facility manager also wanted to revamp organization at the facility. The actions primarily dealt with operator responsibilities and coordination.

The facility manager announced that formic acid denitration, which had run out of control, had been suspended in the unit where the incident had occurred ("the planned corrective measures have therefore been called off."). On the other hand, a number of corrective actions relating to radiological monitoring had been implemented:

– The room's hand-foot monitor had been put back into service;

– An air monitoring system had been set up in the room;

– A hand-clothing monitor had been installed in a room adjoining the room containing the filters.

The lineup error that had resulted in the 2005 incident had been caused in part by poor ergonomics of the workstations of the cabinets on the shielded process line. During the tour, the licensee indicated a number of modifications that had been made:

– The lighting in cabinet C2 had been improved;

– A diagram with color codes indicating the connections between C2 and C1 had been posted near the manipulators on the front panels of both cabinets to allow researchers to match the hoses on the ribbon cables in C1 and C2;

– The connections had been physically modified to make connections between lines for potentially radioactive products and lines for non-radioactive products impossible and thus prevent the possibility of radioactive products flowing into the gloveboxes.

Following the incidents of 2004 and 2005, many actions consisted in creating or modifying the facility's documentation. The facility manager presented three actions regarding the documentary database:

– Rewriting of general instructions to make them as concise as possible (using "is prohibited/is mandatory" phrases);

– Broadening of the modification procedure: "The procedure on technical modifications has been broadened to include no longer just physical modifications, but also modifications of other types, such as process modifications and I&C modifications."

– Creation of a lineup procedure setting out "how to use lineups that jeopardize the integrity of the first containment barrier. Furthermore, each lineup will have to be checked by a fellow researcher or a superior. In each unit there is now a 'lineup record' for listing all lineups and tracking their status so that people know whether a particular lineup is, for example, being used or has been put away."

The facility manager presented four main actions that were designed to improve the setting-up and running of experiments:

– Creation of operating diagram books: "The idea is to create a common language .... The researchers use design documents, which they don't need. What they do need, however, are operating diagrams. The researchers need to be disciplined enough to define operating diagrams that will have to be validated, mounted and checked by the operations manager."

The coordinator told the facility manager that he thought that this would be difficult to implement in the research units. "There are a lot of people involved in the shielded process line. Managing these books is going to be very complicated." The facility manager, who believed in the merits of the books, responded emphatically. "Absolutely not! They will have to shift from construction diagrams to operating diagrams otherwise everyone will have his own diagram. The use of standard diagrams can make the researchers' job easier. We have 200 tanks. That means 200 diagrams will have to be made and placed in each operational unit."

– Creation of operator's manuals for the equipment in the facility;

– Creation of standard diagrams on feeding reagents from the gloveboxes;

– Drafting and implementation of a procedure on erecting vinyl airlocks;

The organizational modifications planned consisted in:

– Beefing up safety function checks: "There are currently three people in the safety function. The aim is to increase that number to five by mid-2006 so that the safety function can perform its duties." (the facility manager);

– Redefining the duties of the operations managers: "They are now appointed by the head of the researchers' department and the facility manager. Their duty is one of safety and security oversight and information within a defined geographic perimeter. Their job description has been rewritten to reflect this. This year we began holding a seminar with all the forty-five or so operations managers to discuss their job more in depth, because they have things to say. The department head has promised to give the operations managers enough time to perform their duties in addition to their research tasks." (the facility manager);

– Strengthening of coordination between licensees and researchers: "In addition to the weekly meeting, there are quarterly meetings between the research departments and the facility manager. There are also interdepartmental meetings, but their frequency has yet to be determined." (the facility manager)

At the end of the presentation, the coordinator told us that he was relatively impressed by all the corrective actions. However, for him, there remained a problem of "managerial legitimacy". "The facility manager's instructions seem good, but getting the researchers to follow them will be a problem. The facility manager doesn't have any managerial legitimacy. The researchers report to other units or departments that have publication objectives and sometimes don't prepare things correctly."

The coordinator subtly entered into this discussion with the facility manager. "How can one be sure that the researchers will do the right thing? The message, which is clear, is yours. But is it accepted in the department?" "It can work if I get the units to agree. I've got a little hammer but big nails! One thing is for sure: it'll take time and there won't be any miracles." (the facility manager)

The safety engineer added the lack of resources to these difficulties. "There's a numbers issue. There are seventy of us in the operations unit, five of whom are in the safety team. There are nearly three hundred researchers. It's not going to be easy spreading the word." (interview of July 7, 2006)

The facility manager did not rise to the bait when, at the end of the meeting, the coordinator asked him the following question: "So, in our opinion, what might help you is if we requested researcher checks?" The facility manager did not appear very impressed by the expert's authority or his ability to change the methods of regulation between it and the heads of the research departments. "You can't ask for more than what people can give!" he said.

2.4. Conclusion: Nature and Outcomes of the Interactions

For the expert, the review went well, probably better than during the same stage of the Minotaure review, when the licensee had refused to allow the human factors expert to interview its operating personnel. This review also had some differences and similarities with the Minotaure review. The information collected by the experts was denser and much more technical than in the previous case. Assisted by the coordinator, the expert eagerly set to work learning about the facility's processes that led to the incidents, somewhat like his colleague had done for the fuel rod fabrication activity at the Minotaure reactor. As we mentioned during the scope definition stage, we also noticed consistencies with the human factors themes assessed, particularly in terms of team coordination and operation set-up.

3. THE DRAFTING STAGE (FEBRUARY–MARCH 2006)

Unlike in the Minotaure case, the human factors expert met with his supervisor between the end of the review and before writing the first version of his contribution. As a result, the supervisor made far fewer changes to the expert's argumentations for each theme.

The contribution's structure was validated during the meeting, which was held in late January. The discussions between the two experts also guided the formation of several opinions. "Our discussion dispelled a doubt: the criterion for initiating risk analyses was not clear .... We saw that we had difficulty seeing the safety team's role in set-up .... [The supervisor] thinks that an ergonomics analysis should be carried out on all the facility's cabinets .... There are a number of things that remain unclear about the RPU's intervention during the 2004 incident." (the expert, interview of February 1, 2006)

At the end of the meeting the expert felt better prepared to draft his contribution. While drafting it, he had no hesitations contacting the coordinator whenever he had doubts or needed his opinion about the draft requests. Despite these precautions, proofreading of the contribution lasted until March 17, the date on which the SEFH's contribution was sent to the generalist engineers. Given that there were not many sticking points, the supervisor's very busy schedule is the most likely explanation for the lengthy proofreading stage. Their discussions focused primarily on the set-up of experiments and more specifically on the lack of risk analysis.

When the supervisor proofread the expert's first draft, he made many recommendations on the document's outline. The second time the supervisor proofread the contribution, he was struck by the following phrase: "the licensee did not directly identify the lack of risk analysis as being a factor [of both incidents]". This phrase was followed by a series of corrective actions proposed by the licensee in relation to the subtheme. Bothered by this, the supervisor wrote the following comment on the expert's second draft: "There's a problem here. The licensee didn't identify the problem yet implemented corrective measures ...". The problem was not resolved until the last minute when, while finalizing the contribution with his supervisor on March 15, 2006, the human factors expert explained that what he wrote was not inconsistent. "We're the ones who find the corrective actions acceptable in relation to the lack of risk analysis even though they don't say that these actions addressed the lack of risk analysis." So the experts inserted the following phrase in the final version of the contribution: "No corrective actions addressing risk analysis were presented in either incident report. The licensee presented the following corrective actions during the technical meetings on December 9, 2005, and January 18, 2006: ...."

The expert also expressed his dissatisfaction by asking two questions that he considered important and that he felt were not answered by the information provided by the licensee: "Does the fact that an activity is included in the safety reference framework guarantee that the activity does not pose any risks that would justify a risk analysis? Do the researchers have sufficiently detailed knowledge of the safety reference framework to allow them to know when they deviate from it and to alert, in accordance with procedure, the operations manager so that he may decide which measures to implement?" (first draft)

Considering the interrogative form unsuitable in an IRSN opinion, the supervisor rewrote the expert's argumentation thus: "The IRSN considers that the current procedure for managing modifications within the facility to be unsatisfactory:

– It falls to the researcher who made a modification to determine, alone and without carrying out a risk analysis, whether such modification may deviate from the facility's safety reference framework.

– The effect of such a procedure is that the safety function is not involved in the first analysis of the consequences of the planned modification. As a result, most modifications made by researchers are not assessed by the safety function. Furthermore, this can lead to the conclusion that a modification is covered by the safety reference framework even before a risk analysis is conducted."

Turning his focus more to the modification procedure, the supervisor asked the expert to explain it in greater detail. The resulting argumentation made a case for a request to update this procedure.

Another matter identified by the expert dealt with the role of the safety engineers in the set-up of experiments. The supervisor polished the expert's argumentation by specifying the measures he would have liked to know of in order to be able to give his opinion on the safety function's resources. The expert's request went along this line. Initially dealing only with the safety function's roles in the set-up of experiments, it was ultimately revised to deal with the resources used by the safety function to perform its duties.

The expert's strategy was thus partially successful. Unlike with the Minotaure case, the supervisor considered that enough data had been collected during the review stage and made corrections to the argumentations in the margins of the contribution.

4. THE TRANSMISSION STAGE (MARCH–AUGUST 2006)

The interactions the expert was able to have with the coordinator facilitated the acceptance of the contribution by the generalist engineers. However, the proofreading stage was not yet finished. After approving the draft report containing the expert's contribution, the coordinator's supervisor sent it to the head of the generalist engineering department. During the approval process, which lasted the entire month of April, the head contacted the human factors experts for their input before finishing the final draft of the report. The experts' work on the case was finished. It now had to be handled by the ASN's inspectors, who referred it to the IRSN. The distribution channel was therefore markedly different than in the previous case.

4.1. The Transmission of the Contribution to the Coordinator

We were able to speak with the coordinator a month after he received the human factors experts' contribution. What follows is a rather lengthy transcription of our discussion. We feel that it clarifies rather well the differences in approach between the generalist engineer and the human factors expert:

"My impression at the start was that the causes of the 2004 incident were technical. For example, there was the evaporator design flaw. But there's also the set-up of the experiment and the RPU's response. This has more to do with organization and what the RPU should have done. The human factors expert's involvement taught me a great deal. I never would have delved so much into things nor in the way he did. When the expert looks for corrective actions, he asks what resources people are going to have in order to do their job. I don't do that. I think that it was a good idea to highlight the issue of workstation ergonomics. The interview portion of the human factors approach brings a lot. The licensee provides important information on resource difficulties and possible pressures. We don't get that kind of information when we do a technical assessment. The facility manager doesn't give it to us. It's the same with ergonomics. Operating difficulties are something that isn't going to be highlighted.

I wondered what kind of background [the expert] had to be able to bring all these matters to light. You see, I use a highly technical approach. There are some questions that I don't ask because I already know the technical answer. I merely read the licensee's report. The licensee explained things whenever the [human factors expert] didn't understand. In the case of ergonomics, I would have gone to see the cabinet but I wouldn't have asked how they fit the hose in place. I wouldn't have considered describing the complexity. But [the expert] said 'I don't understand. Show me.'

I received the opinion on March 17. On March 31, I submitted the draft to the unit head, who proofread it. Now it's with the department head. My idea was to put the incidents in an appendix, but he wants to put them back in the body of the report. My difficulty is that I'm dealing with people who only know the facility through safety documents. I have a hard time getting messages through. It's hard to put the facility's human functioning on paper. It's not that easy. [The expert] noticed things that aren't immediately obvious. I thought the expert's contribution was good and hardly made any changes to it." (the coordinator, interview of April 14, 2006)

4.2. The Proofreading of the Report by the Head of the Generalist Engineering Department

During the few weeks required to draft the final report, the head of the generalist engineering department did not hesitate to bend the institute's proofreading and approval procedures by contacting the experts directly. The comparison of the final report and the human factors contribution revealed a few notable changes.

The head of the generalist engineering department and the two human factors experts knew each other well for having worked together on several previous cases. While proofreading a copy of the draft report sent by the coordinator and his unit head, he noticed a few obscure points. He thought it better to discuss them directly with the experts during a working lunch on May 13, 2006. His first comments were as follows: "The main problem I have when reading the report is that I can't figure out who does what in the risk analysis. I have the impression that the safety function does the inspection but not the analysis."

This led to the following discussion among the three experts:

– The head of the generalist engineering department: Who does the risk analysis and who does the inspection?

– The expert: The safety function does the risk analysis.

– The head of the generalist engineering department: That's odd. What does the researcher do? Does the operations manager supervise?

– The expert: No, it's a researcher who has a few other duties and who in particular monitors and inspects.

– The head of the generalist engineering department: It would seem more natural to me that the researcher start the risk analysis.

– The head of the SEFH: That depends on what you mean by risk analysis. It also depends on the organization.

– The head of the generalist engineering department: I don't understand who does what. Are the tasks the right ones? To me, it's odd that the facility manager is the one who decides on and conducts the safety analyses.

– The head of the SEFH The problem is that there's a big risk of giving the researchers free rein because they don't carry out the risk analysis. The operations manager is the one who is appointed to ensure safety.

– The head of the generalist engineering department: But what does he do regarding safety? Does he enforce the reference framework? Does he do anything else?

– The expert: He also raises safety awareness.

– The head of the SEFH: Because he's in charge and he has to sign his name, he has to make sure that safety requirements are followed.

– The head of the generalist engineering department: The lineup involved in the 2005 incident was not the subject of a modification request. Was it a modification?

– The head of the SEFH: Come on! They added a pump!

– The head of the generalist engineering department: What do they mean by reference framework? Usually it's the safety analysis report and the general operating rules.

– The expert: We haven't really looked at that.

– The head of the generalist engineering department: It'll be worth looking at it closely for the safety review. You have to be sure that there really is a reference framework.

– The department head: You said that the reallocation of roles is a step in the right direction.

– The head of the SEFH: There is another point, which is the revamping of the modification procedure.

– The head of the generalist engineering department: Yes, about that: I've got the impression that a modification can be anything. So, if anything is a modification, nothing is a modification! I can understand for a process modification, but not for an equipment modification. If a risk analysis is necessary each time someone changes a light bulb ... The field of application looked very vast to me.

– The head of the SEFH: We want the operations manager to perform an initial analysis and a copy sent to the safety function.

– The head of the generalist engineering department: OK. I can't tell whether carrying out this request will swamp the operations manager. [Speaking to the expert] Did you meet any? Do they have safety expertise?

– The expert: I see that redefining the task is a good thing, but ensuring it gets done is going to be tricky!"

Following the discussion a few changes were made to the argumentations and requests expressed by the experts. A intermediate version relatively similar to the final version was sent by the generalist engineers to the licensee in early May.

On May 4 the head of the generalist engineering department contacted the expert to get his opinion on the intermediate version:

– The head of the generalist engineering department: I've added three quarterly meetings on coordination improvement. If I want it to get going, I'll have to send it tonight. I won't have the licensee's feedback, but everything has to be sent to the ASN before the safety review kick-off meeting. I had the report read by the assistant manager, who made a few comments. The difficulty is understanding the role of the safety function engineers and the difference between the work of the operations managers and the safety engineers.

– The expert: Is it normal that we all have spent a lot of time on this case?

– The head of the generalist engineering department: It's not related to this specific case. When you want to know who does what, it's not easy. [The assistant manager] also noticed this problem ....

Two weeks later, the report was finished. What were the main changes made to the experts' contribution? To no surprise, the points about the set-up of the operations received the most changes. One of the first additional requests had to do with a fear expressed by the head of the generalist engineering department during his discussions with the experts (overly broad meaning of "modifications"): "The licensee has extended the scope of [the modification procedure] .... Although this change is a noteworthy improvement, the IRSN considers that this new wording is very general and must be clarified. Operating experience feedback from the incidents that occurred at the facility shows that the process modifications generally are not perceived by the researchers as being likely to have an impact on safety. The procedure must therefore be modified immediately so as to explicitly clarify that any modification of a process used at the facility must be documented on a modification request form."

Furthermore, a distinction was made regarding the set-up of associated operations: "The operating experience feedback shows that operations "associated" with experiments (such as cleaning) are insufficiently considered as requiring special preparation and, in some cases, liable to have an impact on safety at the facility."

These corrections were followed by an additional request that was inspired by a "piece of advice" given by the expert and which had not been framed as a request in the contribution: "The licensee shall adopt measures (promotion of the awareness of the Artémis facility employees, audits, etc.) to ensure that the procedure is correctly followed in all cases requiring it and particularly, as stated above, for process modifications and operations 'associated' with the experiments. The licensee shall, within three months' time, present the aforementioned measures. Compliance with the aforementioned requests may be verified during an upcoming inspection."

Lastly, the request relating to the methods of response used by the RPU's experts was withdrawn. To conclude, it should be noted that the generalist engineers influenced the wording of the requests by stressing the safety review that they were preparing.

As in the previous case, the transmission of the SEFH's contribution was followed by a flurry of interactions consisting primarily of collaborations between generalist and specialist engineers. These interactions made it possible to flesh out the points that the head of the generalist engineering department had found obscure. They followed a lengthy proofreading process within the SEFH and, in the human factors expert's opinion, were costly.

One difference with the previous cases deserves mention: the licensee was conspicuously absent from the transmission stage. The generalist engineers had provided the licensee with a copy of the report in order to get its feedback, but there is no formal record of the interaction.

The final report was sent to the ASN's regional division on May 17, 2005. Until the moment when the ASN's inspectors sent the follow-up letter, the licensee did not officially have any say in the assessment's findings. Unlike with the preceding assessment, no open debate was scheduled.

4.3. The End of the Assessment

In early July 2006, we met with two inspectors from the ASN's regional division who were in charge of inspecting the Artémis facility. A few days before receiving the IRSN's report, both inspectors had conducted a human factors inspection at Artémis. During our meeting, they outlined their inspection as follows: "The approach at Artémis is sporadic. Nothing is done. No one knows about the CEA's policy. They don't concern themselves with what goes on outside the facility. There's no ergonomics study. The facility manager doesn't know what the researchers do and is unaware of their training. Nowhere in the organizational memos is mention made of human factors being taken into account. No one knows about human factors aspects in matters such as workstation design. Nowhere in the sheets is there mention of getting the operator's opinion. The human factors aspect of anomalies is not analyzed. The facility manager was incapable of showing us any job descriptions. It's unacceptable." (interview of July 6, 2006)

Regarding the IRSN's report, they found that "The assessment follows the same line. To me, the analysis only deals with the last two incidents. But the conclusions are about the same. We're going to have to make additional requests based on the analysis of the two incidents. The scope of our inspection was much broader." (interview of July 6, 2007)

All the requests in the IRSN's report were repeated by the ASN. By the time the follow-up letter arrived on the facility manager's desk, the assessment of the safety review of his facility was already under way. Given the deadlines for the ASN's requests, it was likely that not all the requests would be implemented during the new review. But perhaps the first assessment would impact the second assessment. We will get back to that later on.

5. PROVISIONAL SUMMARY

5.1. The Basic Operations of the Human Factors Assessment

The main force behind the assessment was the ASN's representative, who requested that the assessment include human and organizational factors. Upon taking up the case, the SEFH's expert in charge of laboratories and plants quickly realized that it would entail analyzing events involving an organization of work and technical processes that were complex. His discussions with his supervisor and the coordinator allowed him to define the scope of his contribution. Brought in early on by the expert, the head of the SEFH participated in drawing up a first list of themes.

After reviewing the case provided by the coordinator, the expert prepared two long questionnaires, one for each incident. Used initially during working sessions with the coordinator, these questionnaires allowed the expert to understand certain crucial technical and organizational aspects at the facility. The review stage went smoothly all the way to the end of the assessment. As per his request, the expert was allowed to tour the facility twice, interview an operator and obtain a copy of an operations document. The expert was accompanied by the coordinator during the tour and appreciated the latter's contributions during the technical meetings. The licensee's answers to the expert's questions during these meetings allowed the expert to reconstruct the timelines that led up to the incidents and identify their causes. To do this, he also used the work of the licensee, which had developed an action plan that was being implemented. In addition to "purely technical" factors, human and organizational factors contributed to the problems: poor activity planning; poor workstation ergonomics; incomplete operating documentation; inadequate team coordination; inadequate inspection methods. The expert took into account the announced improvements in his contribution and requests.

On the whole, the expert's contribution was well received by the head of the SEFH. However, the section on the set-up of experiments required an overhaul that delayed the transmission of the contribution to the general engineers. The argumentations and draft requests were inserted (virtually) unchanged in the report written by the coordinator and proofread by the unit head. However, the head of the generalist engineering department found some obscure points. Going against standard practice at the institute, he discussed these points with the experts. Their discussions focused mainly on the risk analysis and the modification procedure. He wanted to know what a modification was and how to ensure that it would pose no risks. The experts found it hard to answer these difficult questions, which were all the more difficult as the activities conducted at Artémis were part of research programs and the equipment used had to remain flexible. Furthermore, they also had difficulty evaluating the effectiveness of the organizational measures implemented to improve safety (allocation of resources and distribution of roles between the operations unit, the safety function, the researchers and the operations managers).

Six months after the expert's contribution was sent to the coordinator, the ASN's requests were sent to the licensee, which officially received them. At that moment, another assessment had already begun at Artémis.

5.2. Comparison of the Empirical Data and the Theoretical Models

The comparison of the basic operations in this second case and the assertions of the canonical model lead once again to a significant gap. The collective nature of the assessment and the gradual building of knowledge within the very framework of the assessment process once again questions the canonical model. It also seems that the expert's work did not end when the requests were sent since a second human factors assessment was initiated during the facility safety review.

The assessment seems to be less akin to the procedural model than the previous case. There were fewer milestones and systems surrounding the assessment and none of the three principles (independence, transparency, adversial) was embodied by procedures able to guide the work of the experts.

This assessment therefore invalidates several basic assertions of the canonical model and does not follow those of the procedural model.

It is not easy to view the expert's review under the light of Ouchi's forms of control. Admittedly, one can assert that the initiation of the assessment was conditioned by the facility's poor safety results and thus interpret the expert's entire work as a form of result-based control.

It is harder to interpret the operations of the expert's review as forms of procedure-based control. Where then would the reference framework containing the right procedures to be followed be concealed? The list of barriers to be implemented to ensure safety at the facility? As in the case of Minotaure, it was not written beforehand. Nor did it come out directly through the expert's questions, whose chief aim was to understand the event chains that led to the incidents. Once these chains were put together the expert summoned the barriers that should prevent new incidents: good workstation ergonomics, clearly defined roles and responsibilities, adequate documentation, adequate staffing.

The term "procedure-based control" seems inappropriate to describe this operation, which consisted in observing the absence of a barrier once an incident scenario was analyzed. That said, the experts were induced to use in their final recommendation the same objects and actions as in the previous case (e.g. performance of studies, quality of the operating documentation, improvement of incident analyses performed by the licensee), which revealed procedure-based control operations.

The information obtained contains nothing that would seem to call into question the capture behaviors of the expert or the licensee. The two episodes that caught our attention in the previous case did not happen again. The licensee did not discuss the expert's choices of operational methods or get involved in the wording of the expert's requests.

Although it is harder to interpret this case using the theoretical frameworks used, it allows us to become directly aware of the difficulty of the task of anticipating incident scenarios and mastering the complex combination of events that can compromise a facility's safety.

Chapter 5. Management of the Skills of Operating Personnel in Nuclear Power Plants

"What a sad era when it is easier to smash an atom than a prejudice."
Albert Einstein

The third assessment that we will discuss concerned the evaluation of a set of managerial policies implemented in the nuclear power plants operated by EDF. Devised by the central services engineers at EDF's Nuclear Generation Division (NGD), these policies are designed to ensure proper management of the skills of the operating personnel in each of the nineteen plants operated by EDF.

This case contrasts with the previous two in various ways. First of all, it concerns electricity generation plants, not research and development facilities. Secondly, the assessment covered an entire fleet of facilities, not just one facility. Thirdly, the primary contacts of the IRSN's experts were not directly involved in the operation of a facility but were themselves experts working in EDF's central services. Lastly, the assessment was conducted by the human factors experts, who, instead of submitting a contribution to generalist engineers, had to submit a report to the members of the advisory committee.

Although this was the first time that the word "skills" was on the agenda of an IRSN assessment, the training of nuclear power plant operations personnel is a relatively common aspect of assessments and inspections. Indeed, it has been the subject of three assessments and several dozen inspections since 1981. Moreover, experts from the SEFH had taken part in the previous training assessment, held in 1991.

As during the Minotaure assessment, we did not follow the start of the assessment. In order to be able to give an account of the first stages of the assessment, we interviewed the human factors expert who had conducted the assessment and read a number of reports. Starting in March 2005, we began discussing face-to-face with the human factors experts and participating in the various meetings held[175].

1. THE SCOPE DEFINITION STAGE (SEPTEMBER 2004–JANUARY 2005)

The head of the SEFH made the decision to assign the case to a young recruit with a PhD in sociology and who had already conducted assessments of EDF plants. Beginning in October, the recruit devoted all his time to the pre-review analysis. Once finished, the analysis was presented at the various meetings that punctuated the scope definition stage and discussed with EDF's representatives. The ASN's request to the advisory committee marked the end of the scope definition stage.

1.1. First Meeting, First Limitations

The first meeting was held in September. The participants began by introducing themselves then got down to discussing the scope of the assessment. The human factors expert (IRSN coordinator) and his supervisor were present. As we will see, the supervisor would follow the assessment closely. Three members of EDF's Nuclear Generation Division in particular were involved in the assessment. First there was the training budget coordinator, who was the operational coordinator of the assessment. He was often accompanied by the person who coordinated the skills development system project (SDS) at the NGD and whom we will call the SDS coordinator. Lastly, there was a strategic coordinator. He had much experience in dealing with training issues at the NGD and had been EDF's operational coordinator of the 1991 training assessment. In the words of the operational coordinator, "he provides a leadership perspective". In 2004 the strategic coordinator was in charge of a team of engineers from the central services; they would also have a role in the assessment. These engineers were "occupation coordinators" and were primarily in charge of coordinating the management of skills of a particular occupation or group of occupations (such as test technicians or operating personnel) at all nineteen plants. It should also be mentioned that neither the operational, NGD or strategic coordinators nor any of the occupation coordinators were relieved of their functions while working on the assessment.

---

[175] Furthermore, unlike in the previous two cases, we did not directly follow the part of the review during which the experts and external research officers went to the nuclear power plants to collect their data. That said, we did speak with some of them and participated in various meetings during which they shared their observations and analyses.

The ASN was represented in the assessment by a staff member who had been a human factors expert at the IRSN. The experience he had acquired in human factors during his several years of service as a member of the SEFH – he had conducted a preliminary analysis for the assessment held in the late 1990s – made him highly qualified for the job. He was accompanied by an engineer in charge of themes related to PWR reactor operation and human factors.

The presentations given at the meeting were followed by discussions on the scope of the assessment. The IRSN's experts wanted to review the project to predict staff and skills requirements (workforce planning) but EDF's representatives wanted to limit the scope. "As EDF saw it, this forecasting had already been evaluated as part of the skills renewal theme during the assessment of the consequences of plant aging. The assessment therefore wasn't supposed to be limited to skills management. Also, workforce planning is much vaster than skills management. For instance, it concerns policies on matters such as recruitment and outsourcing …. In short, these are matters that EDF considers to be off topic and, furthermore, are no longer coordinated by the same people." (the IRSN coordinator, interview of June 29, 2005)

The IRSN's experts, however, considered that dissociating skills management from workforce planning would be difficult. "Although the negotiations were not easy, we found common ground by agreeing to look at staff forecasting aspects only at site level. In other words, we won't look into the policies devised or the coordination by the central services but may study the matter of how difficulties related to changes in staffing levels are managed in certain areas on the sites. We were afraid how they would react to our insistence on the necessity of doing groundwork, but they didn't object to it." (the IRSN coordinator, interview of June 29, 2005)

This was the first limitation to the scope of the assessment. As we will see later on, EDF's representatives would attempt to limit it further when the expert presented his preliminary analysis to them.

1.2. The Pre-Review Analysis

In order to conduct the preliminary analysis, the expert consulted assessment reports that examined skills management, to wit the three aforementioned reports on training as well as other reports by human factors experts that touched on human factors (in particular a report on the outsourcing of maintenance activities and another on operating crew organization). He also examined a number of policy measures provided by EDF's operational coordinator. In addition, he worked with the SEFH's technical adviser, a former shift supervisor at EDF. Together they selected and analyzed incident reports that implicated a lack of skills. Lastly, he consulted scientific articles and a number of books on sociology, management and ergonomics that dealt with skills. His research allowed him to create an initial analysis grid and an initial list of themes that he would discuss with his peers and present to the representatives of both the ASN and EDF. Guidelines on the review methodology were then discussed at the IRSN.

Following his preliminary investigation and his discussions with his colleagues, the expert developed a model of the skills management system that consisted of a five-stage process. The aim of the review was to be able to collect data in order to evaluate how these stages were addressed by EDF'S skills management process. Accordingly, the experts planned to conduct case studies consisting of interviews and observations of working operators. For each case study, the skills management review had to focus on one occupation. We will use the term "occupation study" in the rest of this chapter.

The analysis of the incident reports showed the relevance of several themes identified during the preliminary investigation. Each of the following items was illustrated in a memo written by the expert in late October by an incident that had occurred at an EDF plant since 1999:

"– *greater consideration of radiological protection* is a significant change that examines the radiological protection skills of those involved;

– the greater emphasis placed on outsourcing by the industrial policy makes *the matter of monitoring and evaluating the skills of outside employees crucial*. Similarly, the use of contractors requires *EDF staff in charge of managing or supervising them to have special skills*;

– skills cannot be reduced to a sum of theoretical knowledge; *know-how acquired through experience must also be considered*;

– *management skills, their identification and their development are an important facet of skills management* in EDF's aim to decentralize decision-making and strengthen involvement on the ground;

– *consideration of collective skills* is paramount once the organization emphasizes the group as the basic unit of its operation and its line of defense."

This memo thus provided information on what the term 'skills' should encompass (theoretical knowledge and know-how) and on the necessity of taking collective skills into account. The importance of considering know-how and the collective nature of skills was also found in the documents read by the expert, who was preparing a review of the scientific literature at the same time.

Initially, the plan was to evaluate the management of a number of skills selected beforehand based on their connection with safety. In the end, the experts opted to perform "occupation studies". But the occupations to be studied had yet to be identified.

The discussions within the department culminated in nuclear power plant operator occupations being classified into three types: facility operation occupations, facility maintenance occupations and support occupations. The expert then established criteria for selecting these occupations. "To be selected, occupations must make it possible to meet a triple requirement: they must be representative of all three occupation groups; they must play a significant role in ensuring safety; and they must answer questions raised either directly or indirectly by the past assessment reports, opinions and event analyses. In order to meet this triple requirement, we propose placing two occupations in each of the three occupation groups: operations manager, lockout/tagout operator, mechanical operations supervisor, skills management officer in a maintenance department of the facility, engineer in the same department, and radiological and fire inspector."

As with the Minotaure case, there was an internal consultation meeting. This one was held on November 8, 2004. Apart from the human factors experts and the representatives of the Reactor Safety Division, few experts attended the meeting. Despite a recommendation by the representative of the operations department to look at the skills involved in operation during incident and accident conditions, these skills were not examined during the assessment. Instead, like contractors' skills, they would be examined in a future assessment. It was also decided to drop the occupation of operations manager since an assessment of the operating organization was planned.

The expert made little progress on the choice of occupations after the internal consultation meeting. He decided to discuss it with EDF's operational coordinator and met with him on November 19, 2004. He presented his project to perform occupation studies as well as the selection criteria adopted. The operational coordinator hoped to gain from the assessment, which he considered to be a "potential source of knowledge"[176]. He readily provided the expert with information and suggested other occupations. "We suggested replacing the operations manager with the technical executive. But the operational coordinator preferred that the study look at the operator. For him, the occupation of technical executive had no future and discussions on the operator's status were under way. We dropped the skills management officer from the list because we wanted to focus on a technical occupation. We agreed together on the electrical technician, particularly since a department manager at the IRSN had suggested it to us. Lastly, we replaced the lockout/tagout operator with the job of first-line manager. A first-line manager plays a key role in skills

---

[176] The IRSN coordinator's words.

management and studying this occupation would help us to investigate managerial skills." (the expert, interview of June 29, 2005)

1.3. The Milestones of the Scope Definition Stage

As we saw in the case of the Minotaure assessment, a number of meetings were set in order to define an assessment. The kick-off meeting was held on December 8, 2004, and was attended by the ASN's representatives. The scope definition meeting was held on January 14, 2005. It was prepared by the coordinators from EDF and the IRSN and attended by the IRSN's human factors experts, who were accompanied by an executive officer of the Reactor Safety Department, and the representatives of both the ASN and EDF.

Little new information was given during the kick-off meeting. The list of six occupations presented elicited no comment from the ASN's representatives

A few days after the kick-off meeting, the human factors expert went back to EDF's offices to prepare the scope definition meeting with the operational coordinator and his colleague, the SDS coordinator. They discussed a number of points together. The occupation studies were the first matter discussed. Given that the forecast schedule left little room, the expert wanted to establish the list quickly: "EDF had set its sights on four occupations and we were aiming for six. We quickly dropped the radiological and fire inspector because, according to the operational coordinator, radiological and fire protection are the job of the mechanical operations supervisor. EDF also wanted to eliminate the first-line manager, but we really didn't want to because leaving it in would enable us to investigate managerial skills. We were possibly willing to drop the technician, but we wanted to wait until the scope definition meeting to do so." (the expert, interview of June 29, 2005)

The discussion then turned to the number of sites: "We thought that they'd give us access to just one site. But EDF ultimately proposed letting us visit two sites per study out of total of three sites. We were very pleased with the proposal. We were still waiting for the list." (the expert, interview of June 29, 2005)

According to the human factors expert, the mention of collective skills particularly elicited a response from EDF. "They proposed to talk about cooperative work situations, which we agreed to do ..." (the expert, interview of June 29, 2005)

The ASN's representatives specified the aims of the assessment at the start of the scope definition meeting. These aims were set out in a letter sent to the director of EDF's Nuclear Generation Division.

> "I asked the advisory committee to give a decision on the following matters in particular:
> – your ability to make sure and demonstrate that you have the necessary skills for operating your reactors in accordance with the safety and radiological protection objectives;
> – the operation of the skills management system that allows you to identify present and future skills needs and provide for them appropriately;
> – the adequacy and suitability of the resources available to the skills management system to allow it to operate efficiently;
> – the suitability of your operating personnel clearance system with regard to safety."

Sidebar 18: The aims of the ASN's request to the advisory committee.

These aims were defined in agreement with the human factors expert. It is worth stating that this was the first time that the term "clearance" appeared[177] whereas it had never been brought up during the kick-off meeting. One bone of contention during the meeting was the list of documents that the experts requested from EDF. "EDF reacted very strongly to our request for documents on recruitment and redeployment [internal mobility]. They felt that it was beyond the assessment's scope. They explained their fear by referring to the difficult labor relations at EDF. We put up a fight. They ended up accepting, but said that they would like us to attend a presentation by their HR Director on labor relations at EDF." (the expert, interview of June 29, 2005)

EDF also considered that the model of the skills management system designed by the experts was ill-suited to the NGD's management processes. In EDF's view, the steps identified were part of processes managed by different departments. They therefore wanted this model to be limited to the processes that they themselves controlled. Now it was the expert's turn to react. Should they change their analysis grid just to further accommodate EDF's organizational requirements? Certainly not. Their decision would be backed by the ASN's representatives.

1.4. Conclusion: Nature and Outcomes of the Interactions

As in the previous cases, we will now list the individuals with whom the human factors expert interacted while leading the assessment, describe his relations with them and mention the outcomes of the scope definition stage:

- The service's technical adviser. *His selection and analysis of the incidents*

---

[177] Clearance "provides formal recognition by the employer of a worker's ability to perform, at a particular facility and for a limited period of time, activities that pose professional risks to himself and his surroundings". Risks are grouped according to three different categories: conventional safety, nuclear safety and radiological protection. (source: EDF document)

allowed the coordinator to define several themes to be reviewed. Furthermore, they developed together the selection criteria for the occupations.

- The coordinator's supervisor. *He advised the coordinator on what resources to use for the review. It was together, and along with other experts in the department, that they came up with the idea of defining the cases based on the occupations. He also participated in the various milestones of the scope definition stage and the negotiations, which were occasionally tense, with EDF's coordinators.*

- The institute's experts. *The coordinator's request for their assistance in selecting the work situations to be analyzed and their participation in the occupation studies during the review stage went largely unheeded by the generalist engineers and other experts at the institute. The studies on the skills of contractors and operations manager were examined and excluded on the basis that they were found to be redundant with current or future assessments. Furthermore, the generalist engineers would perform a quantitative analysis of the links between incidents and skills deficiencies and there was the possibly that the fire and radiological protection experts would write a contribution.*

- The institute's Reactor Safety Division. *The RSD encouraged the experts to not simply perform a documentary study but to collect field data. It also assisted the experts during the negotiations with EDF's coordinators.*

- The ASN's representatives. *During the scope definition meeting they decided in favor of the IRSN (compliance with the analysis grid) or EDF (exclusion of the human resources policies).*

- EDF's coordinators. *The negotiations between IRSN and EDF's representatives during the scope definition stage contributed to changing the themes and review resources that had been proposed by the experts. The scope of the assessment was thus limited. In particular, human resources policies would be considered as contextual elements and the experts would have to use caution when discussing collective skills. The number and range of occupations to be studied, and thus the scope of the data collection, were also discussed with the licensee. That said, the human factors experts identified the beneficial role of the operational coordinator in the assessment: whereas they thought that only one site would be opened to them, EDF's coordinators promised to let them tour three plants.*

As previously, the outcomes of the scope definition stage were the result of many interactions. There were many more exchanges between the experts and the licensee during this stage of the assessment than in the other two cases.

The review of the scientific literature, a variation specific to this assessment, is also worth mentioning particularly on account of the expert's profile.

2. THE REVIEW STAGE (FEBRUARY–AUGUST 2005)

While waiting to receive the documents from EDF, the IRSN coordinator had to move fast to draw up specifications for an invitation to tender designed to expand his team. This because two studies were to be performed by the SEFH's experts and the three others were to be outsourced. Before going to the sites, each expert or group of experts in charge of an occupation study would also meet with the EDF occupation coordinator for their study. Meetings were also held in early April. "On the 4th we went to a training center, where we meet with trainers and managers. Between the 4th and the 18th, we went around the three sites to introduce ourselves. The protocol negotiated with the central services in February was discussed at each site." (the IRSN coordinator, interview of June 29, 2005)

2.1. The Occupation Studies

The occupation studies started on April 18 and ended in early June. According to the IRSN coordinator, this important stage in the data collection process would consist of around one hundred interviews and some forty observations. Each research officer would spend a week at each of their assigned two sites. At the request of EDF's operational coordinator, the IRSN coordinator drafted a research protocol to be followed by the experts' correspondents while on the sites (cf. Sidebar 19).

The interviews

The interviews shall be conducted in a room free of noise and distractions so as to ensure confidentiality and avoid any interruptions. The interviews must be conducted in a context of mutual trust between the interviewer and the interviewee. A dozen people in each field shall be interviewed. The expert shall begin each interview by introducing himself and stating the interview's reasons and aims as well as its ethical guidelines. Each individual interview shall last between one and two hours.

***In situ* observations**

Field observations shall take place over a period four half-days. The expert shall make sure that his presence disrupts the activity being observed as little as possible. The expert's objective is better grasp the skills actually used to perform the activity being observed. Under no circumstances is he to assess worker skills. Informal discussions may take place during these observations.

Validation of collected data

After each round of data collection, the expert shall have the persons interviewed or observed confirm the data or shall reprocess the data a first time.

Debriefing

Following validation, at the end of the observation period, a debriefing on the data collected shall be held by the site correspondent in the presence of the occupation coordinator. During this debriefing, certain information obtained during the interviews or observations may be clarified, confirmed or refuted.

Ethical rules

Individuals interviewed or observed shall be volunteers;

Their anonymity shall be maintained;

No judgment may be passed on individuals;

Only the data required for the review may be sought and processed.

Sidebar 19: Excerpt from the on-site observation/interview protocol for the experts.

To enable the research officers to prepare for their tasks, the coordinator provided them with a ten-page document titled "Interview Targets". Each interview was conducted using the "360-degree interview" technique (see Sidebar 20).

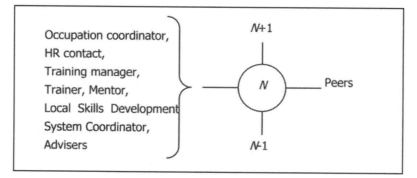

Sidebar 20: Diagram showing the group of individuals to be interviewed when studying occupation N*.*

In the document, which served as a guide for the research officers, the coordinator suggested questions that should be asked and identified those that he considered to be essential (cf. Sidebar 21).

Step one: definition and frameworks

Consideration of non-technical skills

Step two: determine the need

Local implementation of the project to predict staff and skills requirements

Step three: identify the possible solutions

Choice (decision) between the possible answers to a skills need

Step four: implement the solution(s)

The specific skills of trainers, mentors and recruiters

Step five: evaluate

Evaluate skills in the field

Questions common to all the steps

Is the difference between the two sites legitimate, justifiable and problematic?

Integration of experience feedback

National/local structure

Sidebar 21: The questions the coordinator considered to be essential.

Upon their return, we were able to speak very openly with the human factors experts in charge of studying the occupations of test technician and control operator. On the following pages we provide some information on how the experts conducted their studies, the welcome they received and their initial impressions.

First, here are the general impressions of the SEFH expert in charge of studying the occupation of test technician:

"I noticed significant differences between both sites because one of them was in an outage[178], a phase during which many tests have to be conducted. An example of a test is connecting a measuring instrument to an electrical panel in the machine room next to the control room. The weeks I spent on site went well. The management was a little concerned, but the technicians have nothing to hide." (interview of May 17, 2005)

He also provided us with a few details on how his week on the site was organized and briefly spoke about his interviews and observations:

"On Monday morning I interviewed the head of the testing section. ($N + 1$). Then I had lunch with the technicians. In the afternoon I interviewed a beginner test technician.

On Tuesday I observed a beginner test technician with his supervisor. In the evening I interviewed the supervisor.

---

[178] A refueling period. Many maintenance operations are performed during an outage.

On Wednesday morning I interviewed a highly experienced technician, an experienced technician (6 years in the field) and an assistant section manager.

On Thursday I observed an experienced technician.

Friday morning was devoted to feedback and a discussion with the occupation coordinator.

My interviews and observations also covered the area of clearance management. There are four levels of nuclear security clearance across 14 areas of activity for test technicians. I found that two different policies were in use on both sites." (interview of May 17, 2005)

We also spoke with the human factors expert in charge of studying the occupation of control operators. He provided us with some information about the job. "The control operator monitors and controls the facility. There was a nationwide plan to make operators executives, that is, people with managerial skills. To cooperate, I immediately addressed people on a first-name basis. Eliminating formality helps me adopt an attitude that is not that of an expert/inspector." (interview of May 17, 2005)

The expert also identified a matter that he found problematic and which, as we will see, would be raised in subsequent discussions. "One of the main problems I identified is that the operators no longer have time for training courses. This leads to another problem: because they know they don't have time for training, they don't feel that it's worth the trouble to say what training they need." (interview of May 17, 2005)

A first progress meeting was held by the coordinator on May 23. The discussions between the experts are quoted below. The coordinator opened the meeting with the following statement: "Here's a summary of my meeting with [EDF's operational coordinator]. He made four comments that are fairly negative. 1) He found the mean age of the experts to be very young and even too young. 2) Some areas were not dealt with in sufficient detail and not all the questions were asked. 3) He cautions us against straying from the focus of the assessment, which is skills management, not skills or human resources. And last, he suggests not losing sight of what we're here for. Our job isn't to give EDF advice or say what should be done."

During this first progress meeting, the expert in charge of studying the occupation of test technician emphasized the close connection believed to exist between the skills management process and clearance management. In his view, it is an advantage for identifying skills and acquisition methods and scheduling refresher training. On the other hand, he noticed that the transfer of skills from older to younger generations was not easy. "The senior workers I interviewed on one site told me that their younger counterparts are less motivated to learn field skills because they are more interested in management positions. On the other site, the younger workers told me that the older workers were not interested in getting involved in their training." The expert also noticed a "lack of clarification and formalization of methods for acquiring non-technical skills". The next day, the coordinator interpreted the good management of skills observed as a consequence of the occupation's stability: "The occupation of technician is one that EDF typically masters. This confirms a theory found in the literature that management systems are well implemented and even crucial in a stable environment such as the occupation of technician." (interview of May 24, 2005)

The expert in charge of studying the occupation of control operator had identified "a saturation point, in terms of time, in the skills maintenance system": "This bottleneck weakens the entire system, particularly the needs identification stage. I also observed difficulties in the evaluation stage; the peers don't like to evaluate. I also identified gaps in the traceability of routine actions, such as tutoring and mentoring. There's also a problem regarding the recognition of the occupation of control operator trainer and a problem of trainer staffing levels. The positive side is that they are aware of these problems." (meeting of June 8, 2005). The coordinator cautioned the expert: "Careful. Make recommendations about skills management, not about a particular occupation." (meeting of June 8, 2005)

As for the three other studies, which is discussed in the following paragraphs, the coordinator had a few concerns "particularly because of the research officers' inexperience with the relational context specific to the advisory committee." (interview of May 24, 2005). Regarding the occupation of core operations engineer, he considered that the "study is frustrated from the first stage of the process. The occupation of engineer looks to be very complex and the necessary skills appear to be hard to formalize. For that matter, the literature says that management systems only complicate and adversely affect highly unstable occupations. As a result, the process doesn't work. Furthermore, there's a problem in getting a clear idea of what the occupation entails; it's completely different in both cases. The on-site feedback portion didn't go very well. In short, I don't know what we're going to do with this study." (interview of May 24, 2005)

The research officer in charge of studying the occupation of mechanical operations supervisor started his presentation by making an important point. "This isn't an occupation. It's an assignment given to a maintenance technician. It consists in monitoring contractors during an outage, not for technical aspects, but to ensure that they have the proper clearances, comply with labor law and fulfill their contractual obligations. For example, take the case of fire prevention. The mechanical operations supervisor's job is not to make sure that the contractor has installed everything correctly, but to make sure that it has a permit and is thus supposed to know how to properly install everything." (meeting of May 23, 2005). The research officer also identified significant differences between the sites and difficulties in relation to their assignment, or "how to strike the balance between cronyism and policing with contractors". The next day the coordinator commented on the study: "On the whole, for someone who isn't really familiar with the institutional context of the safety assessment, I think that he did a good job. I also noticed a problem with the identification of skills. Because monitoring is viewed as an assignment and not an occupation, it is identified as an actual skill. The skills required to do it are not identified." (interview of May 24, 2005)

The research officers in charge of the studying the occupation of first-line manager presented a few of their observations during a second progress meeting, held on June 8. "We noted two types of career paths. There's the employee who has built his career at the plant by climbing the ladder one rung at a time. He makes a rather low wage, has strong technical expertise and is of a certain age. Around the country, this profile is tending to be replaced by that of the young engineer. Regarding training, we noted that there was nothing about computer tools. On the positive side, we saw that management both takes an interest and is involved in skills management. Interviewees told us that there were local and national 'MPL networks' that were useful and used. On the negative side, we found that the tools used to identify skills, particularly the skills map, are not suited."

2.2. The Review Milestones

The purpose of the second working meeting of the research officers was to prepare a feedback meeting, scheduled for that afternoon, with EDF's representatives. Two days later, on June 10, a halfway meeting was held in the IRSN's offices. Lastly, a meeting was held on June 13 between the IRSN's human factors experts and the ASN's representatives to review how the progress of the assessment.

The feedback meeting at EDF

The atmosphere during the meeting was relatively tense. On one side of the table were the research officers, the coordinator and his supervisor. On the other were EDF's three coordinators and occupation coordinators. The correspondents at each site were on conference call so that they could each speak in turn.

The IRSN coordinator gave his perspective, relatively positive, on skills management at one of the sites. "The potential strong points are a mature system and proactive and wide-ranging skills management. However, there's

one think we'd like to know. Is the extensive equipping a burden for the users?".
To which EDF replied, "We don't understand why you would make such a
comment."

As will be seen, this kind of reaction would be repeated over and over. It
was symbolic of the meeting and the relations that developed between the
experts and EDF's representatives. For example, when the expert in charge of
studying the occupation of test technician voiced his reservations about the
management of non-technical skills, EDF answered by asking, "Beyond
perception, does this have the effect of non-quality of the activity?" As the
department head had anticipated, EDF had an immediate answer when the
word "outsourcing" and its possible effects on technical skills were mentioned.
"This occupation isn't particularly affected by outsourcing." And when the expert
mentioned the lack of recognition of the occupation, EDF responded by saying:
"You have to separate feelings from facts."

When the consultant in charge of studying the occupation of core operations
engineer mentioned staff turnover problems, he was quickly cut off by the
strategic coordinator, who asked the site correspondents to give the "facts on
the turnover rate". Upon hearing the rate given by the correspondents, which
was different from what the research officer had found, the operational
coordinator said: "In terms of facts, I find it a little bit difficult."

The research officers in charge of studying the occupation of first-line
manager then gave their appraisal of the method of obtaining a parameter from
the skills map, a management tool for identifying skills needs. EDF's coordinators
replied by telling them that they hadn't understood it right, saying, "This is a
point that needs to be clarified." And when the research officer mentioned,
during the presentation of the study on the occupation of mechanical operations
supervisor, management's lack of recognition of the occupation, EDF was quick
to respond: "On what basis have you come to this conclusion?"

The expert's explanations about the difficulties reportedly experienced by
training staff led to the following discussion between him and the occupation
coordinator:

– The expert: Are the training staff in touch with reality? One trainer told me
that he's been a trainer for four years and that there's a risk of him knowing less
than the operating crew.

– EDF: But he's in constant contact with the operating crews!

– The expert: It's not a problem of being in contact with people, but with
situations.

The experts left the meeting dissatisfied: "They treated us like we were their
consultants. That's not how it should be. It's vital that everyone's roles be
redefined." They would have the opportunity to make themselves heard during

the halfway meeting held two days later at the IRSN's offices in the presence of the ASN's representatives.

The halfway meeting of June 10

The meeting began with a comment by the head of the SEFH, who lamented the atmosphere of the feedback meeting with EDF. "I want to tell you about the team's dissatisfaction. There were two kinds of problems. The first is related to the definition of the meeting's objectives and the resources you had provided to us. The site workers couldn't even see our presentations. The second is your unconventional attitude during the meeting, which prevented us from discussing technical aspects."

EDF downplayed what it considered to be qualms. The floor was then given to the IRSN coordinator, who gave his assessment of the data collection process: "Altogether, around 90 interviews and 60 observations took place during the seven weeks of data collection. The overall view is positive .... There was mention that EDF was supposed to provide a summary of the national policy. We have yet to receive it." EDF pledged to provide the summary by July 15, adding that the entire approach had nonetheless been presented to the experts.

The IRSN coordinator then presented the points that he wanted to examine in detail during the analysis stage. These points are given in Sidebar 22 along with a few of the reactions of the meeting's participants:

> **1) The assessments** (in work situations; following professional development actions; of employees and the investigators): "We're now fully within the scope of the matter" (EDF)
>
> **2) The relationship between individual and distributed skills**: "Our definition of skill is the ability to make use of what is around you. It seems to me that both we and the sites have given you plenty of information about this." (EDF)
>
> **3) The link between recognition, motivation and skills**: EDF did not dispute the value of this subject but said "when you began talking about motivation, the conversation branches off into subjects such as wage policies, which are interesting, but off the subject." (EDF) "We'll point out the problem but won't trace it back to its sources. We'll stick with 'how skills management integrates these aspects'." (Head of the SEFH)

> **4) The relationship between the skills management tools and the occupations** (adaptation of the tools to changes in the occupations; adaptation of the tools to the specificities of the occupations; dynamics of the tools and situations): EDF drew IRSN's attention to the fact that all the work situations, from the oldest to the newest, were growing increasingly instable. "It occurs in all aspects of management. Tools that destroy meaning are as old as the hills. It also occurs in lots of other areas." (EDF)

Sidebar 22: The four areas of in-depth analysis proposed by the coordinator and commented during the meeting of June 10.

A representative of the ASN then took the floor to comment on several exchanges between the experts and the licensee. "1) We, too, emphasize what's just been said. To us, it's important that the information that's been collected be confirmed. It would be a shame if the observations were called into question during the meeting of the advisory committee. 2) I attach a lot of importance on the written word. It's much better when a licensee is able to write out its policy. That's why I urge EDF to send the written summary to the IRSN by July 15. 3) The advisory committee can't be asked to give an opinion on labor situations within a utility or to set it against a backdrop. This isn't about describing the operators' doubts. That's not the object here.

After stating their expectations and providing this clarification, the two ASN representatives left the meeting so that the experts and EDF's representatives could talk.

The discussion was meant to be more constructive and the participants explored a number of points in detail:

– EDF: Regarding the mechanical operations supervisor, a lot was said about the assignment but not about skills management.

– The head of the SEFH: That's right. This occupation will enable us to think about the assignment/skills management link as well as supervision skills.

– EDF: The policeman/crony balance is interesting, but needs to be worked on.

– The expert in charge of the control operator study: Regarding the control operator, we've set a date with the occupation coordinator to discuss the map, clearances and training courses.

– EDF: There are some good questions regarding the first-line manager. Both the map and aspects about the professional development path need to be rediscussed.

Progress report with the ASN

Three days later, the experts met again with the ASN's representatives to report on their progress. One of the ASN's representatives went back over the clarifications he had to make during the halfway meeting, particularly on the labor aspects, which he viewed as beyond the assessment's scope:

– The representative of the ASN: Did the interaction with labor issues crop up during the discussions?

– The IRSN coordinator: Yes, particularly at one site where morale is very low. Some people there voiced worker demands. But we didn't perceive it to be a problem.

– The head of the SEFH: The avenues to be pursued aren't contaminated by those aspects.

– The representative of the ASN: My concern is that the discussions before the advisory committee should not be affected by these aspects.

– The IRSN coordinator: We don't see the review the same way EDF does. They want avenues right away and "audit/inspection" type feedback. However, we do agree with them about the additional information.

The meeting then turned to the upcoming program and the presentation before the advisory committee. Before ending the meeting, the ASN's representative re-read the objectives of the request (cf. Sidebar 18, p. 167) in order to assess the progress made:

– The representative of the ASN: The first point is couched in rather process-focused terms to avoid asking EDF if they employ skilled people.

– The head of the SEFH: It's the most difficult point. EDF might respond by saying "the system creates skills", but whether people are skilled is not known. We're a little in the dark about the aspect of skills effectiveness.

– The representative of the ASN: That's not an issue. The question is "is EDF able to demonstrate that the specifications are good?" We wrote "EDF's ability to demonstrate", not "to have". In other words, say EDF draws up a bad frame of reference for an occupation. Will you notice it?

– The representative of the DSR: We'll point out shortcomings, but I'm not sure we'll get an answer.

– The coordinator: We could, for example, say that the distributed skill isn't covered by the frame of reference.

– The representatives of the ASN and the DSR: That's a very good point.

– The coordinator: We've got a few things on resource allocation.

– The representative of the ASN: Clearance is important for us. Have you got anything on that or on the "suitability with regard to safety"? Are you able to come to a decision about it?

– The head of the SEFH: That's something that we have to look into more closely.

The human factors experts left the meeting relatively satisfied about these last two milestones. "The discussions on the 10th were more polite .... EDF's attitude towards our comments and questions was much more balanced. Basically, there was less 'No, we disagree with your point of view' and more 'We don't deny that ... but perhaps ...'. The meeting on the 13th provided the opportunity to rediscuss everything with the ASN in a calm and constructive manner." (email sent to the research officers by the coordinator on the afternoon of June 13, 2005)

2.3. The Occupation Study Reports

The five occupation study reports were submitted in July and August. After being closely read by the coordinator, they were forwarded to EDF's representatives in September.

In June, the coordinator sent recommendations to the research officers on several occasions. In one e-mail, he emphasized the requirement for support. "Include details that underpin your statements (document excerpts, verbatim transcripts, observation reports, etc.). If the excerpts bog down the body of the report, you could perhaps place them in an appendix. It's your decision. The important thing is that, when the five studies are combined, we have tangible evidence on which to base our arguments. In short, if the Shmurtz argument is based on two excerpts from study 1, both have to be retranscribed and you have to say who said what and on what site. This way, we'll be able to see if the argument is specific to a site or occupation, is related to a specific status or is more general in scope. We'll see later on if we should keep names anonymous or not." (email of June 27, 2005). He also provided the research officers with bibliographic references, such as an IAEA document on skills.

In late August, the reports were practically finished.

During an informal meeting between the head of the SEFH, the IRSN coordinator and the two research officers on August 29, the human factors experts raised the question of where to send the occupation studies and why:

– The IRSN coordinator: What we want is for the sites to provide frank and honest feedback .... The local and central levels disagree when apart but agree when in meetings together. We want the sites to confirm the verbatim transcripts.

– The department head: Careful there. Confirming a verbatim transcript is different from confirming a verbatim transcript in an argumentation. We can't send them directly to the sites. We have to send them to the EDF coordinators and specify that we want feedback from the sites' departments.

The next day, the IRSN coordinator sent an email to EDF's operational coordinator. "We are planning to send each occupation study to the relevant occupation coordinator and one or two people on each site in order to get their feedback. The individuals in question have been identified by the authors as playing key roles in the studies. The studies will be sent in five different messages, with both you and [EDF's strategic coordinator] being cc'd in each message." (email of August 30, 2005)

In an interview a few weeks later, the IRSN coordinator recounted the last discussion, rather tense, he had had with EDF's operational coordinator. "According to him, it was no longer possible to interview anyone on the site because the collection stage was over. He considered that it was not for the sites' employees to confirm the verbatim transcripts because they were not authorized to speak on behalf of EDF's Nuclear Generation Division. In his view, we were deviating from the protocol. I discussed it with the [department head], who spoke to [the DSR] and [the ASN's representative]. He didn't think that that was right and was willing to do something. But, a few days later, [EDF] called me back to say that, in the end, they agreed to allow us to ask more questions to the sites' employees although confirmation had to take place centrally ...." (meeting of September 20, 2005)

3. THE DRAFTING STAGE (SEPTEMBER 2005–FEBRUARY 2006)

In the previous two cases, we were surprised by the course of the drafting stage, which could not be boiled down to the finalization of an individual analysis followed by approval by management. The numerous interactions with the department head had considerably modified the draft of the contribution to the Minotaure assessment. During the assessment of the Artémis incidents, the human factors expert had collaborated with both his supervisor and the coordinator during the finalization of his argumentations. In the present case, the task was clearly even more complicated. The first version of the coordinator's report was the result of collaboration between the research officers under his coordination. The final version, which was sent to the experts on the advisory committee, included feedback from the department head and members of the IRSN's Reactor Safety Division. It also contained critiques from EDF's representatives, who had been contacted several times to comment on the interim versions of the argumentations and draft recommendations.

3.1. One Analysis Stage

A meeting was held between the coordinator, his supervisor and the research officers on October 11. The department head opened the meeting by stating that "It is important that we issue occupation studies, otherwise EDF will say that the issues we're talking about are specific. Furthermore, our concern is the skills management process, not the skills for a specific safety-sensitive activity." The ensuing discussions were organized by theme and supplemented by a table summarizing the interim findings, which are listed in the following sidebars.

The definition and listing of skills

> There are local and national frameworks (or documents that serve as such). Not all the frameworks are usable, complete, finished, established or updated, but the basis is there.

The coordinator was relatively satisfied with EDF's level. "To be honest, I wasn't expecting to see non-technical skills identified anywhere."

> Non-technical skills have been identified but more remains to be done (specify the acquisition methods, go deeper into their establishment, explain certain skills and their implementation, etc.). A good start, but must do better.

The research officers commented on this passage:

– A research officer: But can we do better than them? And if so, how?

– Another research officer: Is it the IRSN's job to advise anything? When I tried to advise EDF, they told me that it wasn't my job to do so.

The research officer in charge of studying the mechanical operations supervisor also considered that the skills frameworks are established on a national level and "bear no connection with the real world".

The department head continued: "In order to make the frameworks better, you have to start with how they are used. 1) Examine each occupation and ask yourself what its key skills are. 2) Are they listed in the national document? 3) Do the sites consider that the national document provides anything?"

The research officer replied by saying that "The sites tend to say that the national document doesn't do anything special. But if they said why ...."

A discussion between the coordinator and his supervisor then followed:

– The department head: Regarding non-technical skills, they're going to tell us that there are no problems and that the occupation coordinators form a network able to see everything that happens in the field. So we have to be ready.

– The coordinator: We have to identify the points that we're going to grapple over.

– The department head: For the moment I don't think that they're being very clear about the framework-task relationship. Changes have to be taken into consideration as they happen. EDF should be asked what they think.

The coordinator then mentioned the tensions between the "incompleteness" of a framework and its "usability", which he had observed in particular during his review of the literature:

– A research officer: The problem is that the more you describe your framework, the more you risk having people doing their jobs with their noses buried in the framework and no longer asking questions.

– The department head: Yes. I can imagine an expert on the advisory committee asking something like "Have you identified the risk of sticking too much to the framework and no longer the occupation?".

The identification of skills needs

> Needs for technical skills are relatively well identified. A question remains about the skills targets in the maps.

The values and limits of the maps were briefly discussed during the meeting:

– The coordinator: I think this tool is very good.

– A research officer: Isn't it too abstract for the sites? I'd be inclined to say that they don't need it for operations.

> Needs identification is based chiefly on individual interviews and is converted into maps in the departments. Identification during activity is in its early stages.
> There are some limits to identification (the group, motivation, the organization in the case of monitoring).
> Skills identification is strongly linked to the evaluation arrangements.

A research officer emphasized this last point:

– A research officer: I have a really hard time distinguishing the identification phase from the evaluation phase.

– The coordinator: You're not the only one. They're closely connected in all the case studies.

Clearances

> Clearances (concerning technical skills related to safety) appear to be issued and renewed in an appropriate manner. Nevertheless, there is room to wonder about the occasionally routine nature of clearance. Revoking clearance has serious consequences on employees (particularly regarding advancement).

This theme was then discussed by the participants:

– A research officer: The problem is that people get bonuses when they're granted clearance. As a result, revoking clearance is very tricky.

– The coordinator: I can't see myself defending this idea.

– The research officer: And yet we saw it, didn't we?

– The department head: What would you have had to see to make that statement?

– The coordinator: You've got to be hardened to say that.

The department head then made proposals for further review of the clearance renewal issues. "Maybe we can ask EDF how many clearances are not renewed each year. There are two possible cases: 1) The activity in question is not too safety-sensitive and the operator regularly refreshes his skills. In this case, there is no reason why his clearance would be revoked. 2) Problematic cases, such as group activities where the group can hide individual lack of skills and sensitive activities. What is done in such cases? It should be noted that the ASN already asked the question in 1999. It wants us to take a close look at it. On the day of the advisory committee meeting we can't walk in and say that the question seems interesting."

Distributed skills

> Core issue. All the studies show that the occupation studied requires using skills possessed by other employees (in the department, on the site, at the branch and even in the company). On the one hand, skills management seems to be individualized and thus does not consider the skills networks used by these employees. On the other, metaskills (skills in using other skills, knowing where to find knowledge, etc.) are completely disregarded. In addition, the high occupational mobility of these employees makes identifying these distributed skills and their method of distribution crucial.

A research officer wondered how it could be possible for the evaluation to be anything but individual. The coordinator replied by mentioning occupations for which collective evaluations are conducted, citing the cases of air traffic controllers and sales executives. He also provided a difference between collective skills and distributed skills. "A distributed skill is not a collective skill. The first is linked with a network while the second is linked with a group."

The coordinator spoke about the meeting's value during an interview a few days later. "I enjoyed the discussions on updating the frameworks. We also highlighted the fact that better consideration could be paid to the group but I don't know if that will hold water before the advisory committee. At least two reports stress near-routine renewal. But, contrary to what [a research officer] said, nothing was observed; there are only verbatim transcripts. I don't want to draw the connection with wages. That seems unsound and hazardous to me. We'll have to rediscuss it together." (interview of October 28, 2005)

3.2. Further Review

The research officers' final interviews went smoothly. During one of them, at which we were not present, the coordinator even congratulated the research officer. "A lot of people were surprised by the strength of the analysis and conclusions. I find your report to be of very good quality and I use it." (conference call of October 24, 2005). The occupation coordinators merely made a few clarifications or minor corrections. Although they occasionally disagreed with the research officers' analyses, the veracity of the facts contained in the reports was no longer questioned.

While completing the first version of his report, the IRSN coordinator arranged to meet with the operational coordinator and the SDS coordinator. Two occupation coordinators were also present at this meeting. The list of questions that would be discussed during the meeting was sent beforehand to the operational coordinator. The first question dealt with clearance management:

> Compared with the number of renewed clearances, how many clearances have been revoked or suspended on grounds other than medical (by occupation and by site over the past 4–5 years)?

The responses of EDF's representatives are provided below:

"That's an interesting question, but we're unable to answer that. It's not a management indicator. Clearance falls within the domain of management. The evaluation policy covers skills acquired through training courses. Post-training feedback from training staff is used in managing clearance. If a person fails the evaluation, it may be because they did not achieve a training objective. The manager then has to decide whether the lapse constitutes grounds for revoking clearance and he has to implement professional development actions to address the area for improvement. A person who does not pass the evaluation does not necessarily have their clearance revoked. It is suspended if a lapse is very significant or when professional development actions do not address the area for improvement. There is a certain confidentiality about lapses and areas for improvement that falls within the scope of the managerial relationship between employees and their supervisors. The clearance of operating personnel is temporarily suspended if they fail the evaluation by a large margin. If they fail by a small margin, efforts are undertaken to remedy the lapse. A record is kept of lapses. Clearance is revoked when a lapse is not corrected. Clearance revocation is a poor indicator in skills management. We wouldn't want to show it."

The next question was about supervision:

> Are there any sites where supervision is an actual job and not just a task? If so, which ones?

"The central services don't have an exhaustive list because the sites are free to organize themselves as they see fit to carry out their activities. That said, there are sites that have created the position of mechanical operations supervisor. But there's no nationwide move to bring this model into general use." (EDF)

The coordinator finalized certain sections of his report using the information obtained during the meeting.

3.3. The Drafting of the Report

As we have seen, the drafting of the report was preceded by an analysis stage involving several experts, including the department head. And once again, the coordinator brought in his expert colleagues and supervisor to proofread certain sections of his report. It can thus be said that this report was more of a collective effort than the reports in the previous assessments.

After completing the report and having it proofread by his supervisor, the coordinator presented his conclusions to the Reactor Safety Division on December 21. A first draft of the report was then sent to EDF's representatives, who provided feedback during a meeting held in early 2006.

Presentation of the report before the Reactor Safety Division

The coordinator was assisted during this working session by his supervisor and two department colleagues, each of whom had conducted an occupation study and had followed the review from its start. The director and two assistant directors of the Reactor Safety Division (RSD) commented on the report during the coordinator's presentation.

The following critique, for which the experts were planning to write a draft recommendation, drew a reaction from the RSD:

– The coordinator: There's a problem with document updates.

– The RSD: Careful here. You stress the operational aspect, which is good. However, if you ask the central services to accelerate the updates, they will just churn out updates and it will no longer be operational!

– An expert: The aim of the recommendation is to improve usability.

– The department head: Thought needs to be given to the role of the national skills identification guides. If they are meant to be valid for five years, their content has to be up suited.

– The RSD: Which is the overriding factor: frequency or updating?

– The department head: We'll have to think about it.

The determination of skills needs also prompted a discussion:

– The coordinator: On the whole, there are many systems for determining skills needs and they seem to work. Nevertheless, we suggest recommending giving employees more say in what skills they need. But the problem is that they no longer want to say what they need. To them, it's pointless because the time they have for training is already set aside for mandatory courses required by the ASN.

– The RSD: How much of professional development training is required by the ASN?

– An expert: The aim definitely is to want to streamline things, to see what is useful and what isn't.

– The RSD: It's a recommendation that sends a message of "let's think!". There needs to be fewer mandatory things and more leeway.

– The coordinator: That's right. The guys no longer ask for anything or give any feedback.

The director voiced reservations about the content of three draft recommendations made for the "evaluation of abilities, skills and actions taken" stage:

– The coordinator: EDF wants its managers to conduct evaluations in the field.

– The RSD: That's great!

– The coordinator: The problem is that it's done very little and it's not outlined. They'll watch, but they don't conduct evaluations.

– The direction: That doesn't strike me as being part of a manager's job. Imagine me arriving at the SEFH and outlining my observations!

– The coordinator: The wording of the draft recommendation is "If EDF wants to conduct continual evaluations in the field, it must give itself the means to do so."

– The RSD: Why not ask the peers to do the evaluations? Are three draft recommendations necessary for this point?

– The department head: We'll think it over.

The discussions on clearance management mainly focused on clearance renewal, which was the subject of one draft recommendation.

– The RSD: If the renewal system isn't good, maybe we could imagine a point-based clearance system.

– The expert: Clearance was meant to be a barrier of defense. It's like if you stood by and let a machine not do what it's supposed to.

– The department head: But even if they don't revoke a person's clearance, they don't just sit around and do nothing.

– The RSD: How is it done at the SNCF?

The experts did not answer the question. Instead, they mentioned the situation in the USA, where licenses are issued by the NRC.

At the end of the meeting, the directors congratulated the coordinator on his work. There was still work to be done, however, for EDF's coordinators would soon have their turn to speak and react to the experts' conclusions.

The reactions of EDF's representatives

On January 6, 2006, the SEFH's human factors experts and the coordinators and occupation coordinators from EDF's Nuclear Generation Division (NGD) participated at the preliminary preparatory meeting in EDF's offices. The meeting was proposed by the department head. This meeting, as well as a final review meeting attended by the various coordinators a few days later, led to a number of rephrasings, some of which were extensive. The statements are shown in the same type of diagram used in the Minotaure case. Statements formulated prior to these discussions are preceded by the letter "A" while those formulated after these discussions are preceded by the letter "B". Draft recommendations that the experts considered clarifying after the preliminary preparatory meeting are shown in a black box.

The first round of discussions – which covered the record of the "skills management" theme, which was the subject of several inspections – suggested that the meeting was going to be a difficult one.

– EDF: There is an inspection that was left out.

– The IRSN coordinator: It's not exhaustive.

– EDF: But the record has to be complete. It's too old and gives the impression that none of the concerns have been addressed.

– The department head: Straight off the bat we're not going to make an exhaustive record. We can discuss the facts, but we stand by our choice.

EDF's representatives also discussed the incidents chosen by the experts to illustrate the importance of skills in ensuring plant safety. The discussion then moved to the definition of the term "skill".

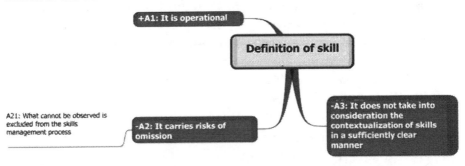

The human factors experts' appreciation of the NGD's definition of "skill" elicited a strong reaction from the division's representatives.

– EDF: First you note that we use observable and measurable elements and place "interpersonal skills", then you say that that's a simplistic view.

– The IRSN coordinator: The skills are taken out of context.

– EDF: That's not true. One isn't skilled in theory; one is skilled for a specific activity. *[reading the report]* "Skills in situation". That's in the field, not the lab.

All right, the word "recommends" doesn't appear, but that's not a major issue. However, the people who will read it will think that at EDF we don't take context into account and that we're not very bright ....

– The department head: Perhaps a few expressions, such as "simplistic view", need rephrasing. But your definitions give the impression that the resources are individual and not shared. Are you shocked by the second point, the one on the restriction of skills to those observed?

– EDF: That's short of crass! It's like an indicator. You have to add that skills aren't measured solely during field observations. There are the results ....

– A human factors expert: "Observable" doesn't mean just "field observations".

The final version contained a few rephrasings and the reference to EDF's "simplistic view" of skill was left out.

The discussion then moved to the first stage of the standard skills management process: the registry.

The NGD's representatives reacted to a draft request to simplify and streamline the documents used to list and identify skills.

– The department head: The idea is "it's very complicated, so streamline it". The idea is to simplify things. Do you agree?

– EDF: There are a two things. First, there is harmonization, which was done. Maybe it is time to streamline.

– The department head: In that case, we'll write the following: "EDF has harmonized and is currently streamlining".

– EDF: Yes, a simplification.

– The department head: The risk is that in 2 or 3 years you'll be asked where you're at.

They therefore agreed on the principle of a draft recommendation that would be "accepted" by EDF and become a stand/action[179].

---

[179]A "stand/action" is to EDF what a "commitment" is to the CEA (cf. Chapter 3, p. 113).

The initial version of the report contained a draft request on the updating of the national frameworks. As you will remember, this theme drew criticism from the DSR's management. It was also discussed with EDF's coordinators and ultimately rephrased.

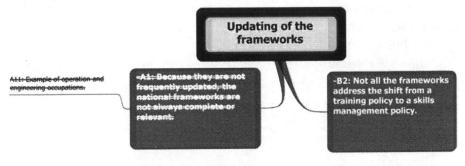

The relevance of argument −A1 was discussed. Ultimately, it was withdrawn and did not make it into the final version.

− EDF: If you look at key activities, things don't considerably change all that often. We think that the form adopted suits how changes occur in the occupations. The risk is that the documents might be off the mark. The aim isn't to do updates once changes occur.

− The IRSN coordinator: Actually, there are two parts. A 'frequency' part and a 'criteria' part. It seems that some changes don't lead to updates.

− EDF: You're touching on the national/local structure. There are plenty of examples that show that you proceed based on the site's history and labor relations. We called that a "framework", but it's not prescriptive. Take the 1997 guide on operation occupations. The core of the occupation doesn't change much. What we end up with is a national guide tailored to local conditions.

− The department head: The issue of updating is related to the clarification of the status of the document. Depending on its purpose, the document may have a prescriptive function on the core of the occupation and a support function on the rest.

− EDF: Aren't you talking about management style here? Some want recommendations while others want degrees of leeway. A recommendation will influence our management style, so you need to be careful with the phrasing.

− The department head: Yes, but it bothers me to address that because it's a difficulty.

The theme was rephrased in the final version of the report. All reference to updating was replaced by ensuring consistency, in the event of changes, with the shift from the training policy to a skills management policy.

Two other points on the identification of individual needs assessed by the experts were discussed during the meeting.

The first dealt with the identification of skills by the managers in the field. The experts identified a possible confusion between the task of identification and the task of evaluation. The operational coordinator reacted to this comment:

– EDF: You imagine that it's possible to clarify the identification of needs and the evaluation. But both are done by the same manager!

– The IRSN coordinator: That's the problem!

– EDF: We reject this recommendation. It could lead to a heated debate during the advisory committee meeting!

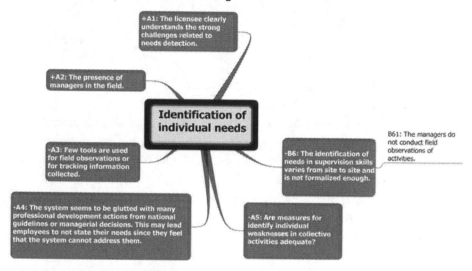

The second point dealt with self-expression of needs (argument -A4). It may be remembered that the officer in charge of studying the occupation of control operator had identified the connection between access to training and self-expression of needs.

– The human factors expert: The system stifles the expression of needs. There was a problem of self-expression at one site.

– EDF: Please, don't say that! The fact is that the simulator and instructor are available. It's one thing when a group of people say "We asked for a simulator but didn't get it." If a person says "I don't dare say what I need," I have no idea what to do. "I don't really want to say what I need" – that's not a positive thing to hear. It's based on statements.

– The department head: There are two points: 1) The effect of censorship on the identification of needs. In principle, there won't be a recommendation on that. 2) The issue related to lack of additional training opportunities.

– EDF: You'll have to prove that point. You'll have to prove that there are problems getting simulators. If you make a recommendation about that, that means that you're expecting something even if you don't provide a solution.

– The human factors expert: What I have in mind are things that are mandatory and which no longer should be. There is some leeway on this aspect.

– EDF: That's not what's written. We're giving thought to refresher training. The aim of the study was to draw general conclusions on skills management. But this is highly focused on just one occupation [operation].

– The human factors expert: But it's not written like that in the report. We're talking about balance here!

The meeting then turned to the following stage of the process, which dealt with the evaluations. The sensitive subject of clearance management, mentioned in the ASN's request, was raised. The ensuing discussion, which may seem rather surprising, illustrates a subtle mechanism of the assessment. The experts had to respond to the request of the ASN, which vets clearance, yet they had little information with which to make a decision. But they were not alone. During the technical meeting on clearance management, the central services stated that they had no information on the renewal and revocation of clearance. However, the experts were convinced that things were there and that they appeared to be sufficient. Given the subject's sensitivity, however, they had to form a conclusion.

– EDF: We were taken aback by the entire paragraph. "The meaning has changed from one of qualification to one of professional recognition." What statements are the basis for this sentence? The texts haven't changed!

– The IRSN coordinator: The starting point isn't a change over time; it's the matter of renewal.

– EDF: I think that the meaning of "clearance" hasn't changed. There is a willingness on the NGD's part for renewal to be made based on proven skills. Since we've raised the evaluation's requirement level, the actions we implement don't call clearance into question.

– The department head: That's what we also want to say to the ASN: there are intermediate things, so don't get carried away with clearance.

– EDF: We agree on that point! The way it's written, it sounds like it has lost its original meaning.

– A human factors expert: But renewal has somehow become an administrative procedure.

– The IRSN coordinator: It no longer provides a trace of skills.

– EDF: ASN's belief that "clearance must be revoked when someone is no longer competent" is wrong. There are many more systems and actions. And when nothing works, we revoke. But we pay attention; we try to identify weak links. If a person may have an effect on the facility, we revoke their clearance .

– The department head: I'll hazard an opinion here. I think that there will be a recommendation on this along the lines of "clarify". The ASN wants clearance for oversight purposes. So, something will have to presented. Each successive line of defense will have to be shown.

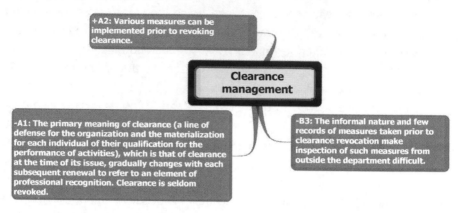

The section that followed dealt with consideration of specific skills. A number of passages of a lengthy discussion on supervision skills prompted a strong reaction from EDF's representatives.

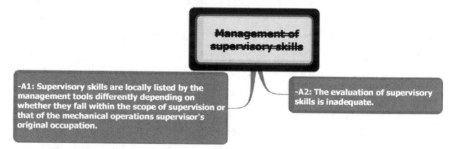

– EDF: You write, "It appears that the professional development of the supervision assignment is currently not clearly defined. There are no national guidelines on the status of the supervisory function." You can't leave that in. There are summaries and information that don't jibe. For example: "a recent assignment". We can't accept these terms.

– The IRSN coordinator: This section is too strong and in the wrong place.

– EDF: What bothers me is that you write extensively about supervision and make a recommendation about it when it isn't even the subject. It's beyond scope.

– The department head: It concludes with observations about recent cross-disciplinary skills.

– EDF: But you make recommendations about the mechanical operations supervisor. These recommendations are beyond scope.

– The IRSN coordinator: There currently is no management of supervisory skills.

– EDF: I disagree. There are frameworks, mentoring sheets and skills evaluations in the field. You say that there are gaps? I'll do everything in my power to show that there aren't any! And then there's the matter of the skills map.

– The IRSN coordinator: Supervision is just one aspect of the map. But is there a map of supervisory skills?

– EDF: No! And neither for verifications or inspections.

– The department head: But the mechanical operations supervisor works full time on an activity.

– EDF: That's the project approach. When a project manager needs someone, that person is chosen by the department head. He knows what the guy's done. It's the same for the heat balance. We don't look at every single skill; what we're interested in is a person's macro-skill! It's the same for managers! We want good managers; we don't look at each individual skill.

The department head showed his satisfaction with the preliminary preparatory meeting several times during the discussions with EDF's coordinators. "Today's meeting served its function as a safety net," he said to the strategic coordinator.

Development of the recommendations

The coordinator included the reactions expressed by EDF's coordinators during the preliminary preparatory meeting and the information collected during the technical meeting of January 11 in a second version of the report. This version was sent, along with the draft recommendations, to the participants of the preparatory meeting.

However, a number of last-minute changes were made. Two days before the preparatory meeting, the experts came together before conferring with their assistant director. The meaning of each verb was discussed and each word was carefully weighed: "Should 'continue' or 'strengthen' be used?" (an expert, meeting of February 1, 2005)

The experts put the recommendation's "quality" into perspective. They developed a strategy by ranking the recommendations and taking the licensee's likely reactions into consideration, such as "That's a bogus recommendation; that's their thing."; "We're not going to fight on that."; "That shouldn't be an issue; they'll propose a stand/action."

As the draft recommendations were read out, the assistant director commented them, saying "I agree with you. That's common sense. Harmonization is necessary!" and "They're going to put up a fight on that!" (meeting of February 1, 2005)

3.4. The Preparatory Meeting

The preparatory meeting was attended by the human factors experts, their assistant director, EDF's coordinators and occupation coordinators, and two ASN representatives who had participated in the previous milestones. The DPN's safety headquarters delegate, who, as usual, would represent EDF before the advisory committee, was also at the meeting, as was the Nuclear Generation Division's human resources director. Lastly, there were also four members of the advisory committee, of whom two were research lecturers in the social sciences[180]. After a quick introduction by the coordinator, the draft recommendations were read out one by one. A transcription of the ensuing discussions is provided on the following pages[181].

The first three draft recommendations dealt with the first stage of the skills management process. The first two — on the updating and streamlining of the skills listing documents — were discussed during the preliminary preparatory meeting. As they had implied during the preliminary preparatory meeting, EDF's representatives accepted the draft recommendations. They thus proposed stands/actions at the end of the meeting.

> **R1.** Insofar as EDF has shifted from a training policy to a skills management policy, the advisory committee recommends that EDF adapt the occupation frameworks and the future professional development frameworks ... and ensure their consistency in the event of changes.

– EDF: We rather agree on this point. We want to do it in the very near future.

– The department head: There's no deadline because there are lots of occupations. However, it would be interesting to give a work schedule if a stand/action is favored. I want to point out that what matters is how things are done on the sites.

– EDF: That's been our concern so far. You noticed that it questions the national/local structure.

– The department head: Is this more of a stand/action?

– EDF: Yes, to the extent that we set milestones for what we intend to do and we take into account the service provided to the sites. So, we will write a stand/action and then see.

---

[180] In the case of human factors assessments, human factors experts who accept the ASN's invitation to participate in the advisory committee meeting are referred to as "advisory committee members".

[181] Only the most-debated draft recommendations are transcribed here.

> **R2.** The advisory committee recommends that EDF simplify the national and local documentation system on skills management so as to facilitate its adoption and utilization ....

– The IRSN coordinator: That ties in with what we just said.

– EDF: Well ... yes! Isn't it the same thing?

– The IRSN coordinator: Not quite. There are perhaps too many tools and systems. The streamlining effort we observed on one of the sites is going in the right direction.

– EDF: We somewhat agree. But be careful. The project conducted on this site was based on a group effort. If a site develops its own management system, we don't want to force it to change it. It's hard to state milestones in an action plan. We can undertake streamlining, but we can't give orders. I have the impression that, at the local level, R2 is a consequence of R1. Couldn't the idea of simplification be added to R1?

– The department head: OK. Propose a joint stand/action.

The third draft recommendation specifically concerned management of the mechanical operations supervisor's skills. The experts expected resistance from EDF's representatives, but were surprised by their reaction. "We're on the same wavelength with this recommendation. We share the same determination to help the sites. We are proposing a stand/action that would go be along the same lines. We will use the best good practices. The strategic importance of this assignment calls for special measures." (EDF)

Three draft recommendations dealt with the following stage of the standard process. The first pertained to the identification arrangements in general. As during the preliminary preparatory meeting, the possibility of equipping managers' "field observations" drew a reaction from EDF's representatives

– The department head: We'd like to know what tools the managers have to identify skills needs in the field. We were left wanting a little bit on that.

– EDF: We really don't understand. Our idea of a manager in the field is not someone with a notebook in his hand. A manager knows what he's supposed to observe when in the field.

– The IRSN coordinator: I propose deleting R4 and talking about this theme with R12.

The following draft recommendation also dealt specifically with the activity of supervision. At the end of the discussion, only a partial decision had been made on its status.

> **R5.** The advisory committee recommends that EDF clarify the arrangements to be implemented for identifying the skills needs of the mechanical operations supervisors in consideration of the performance of their work outside the department, in an intermittent manner ...

– EDF: It's in part similar to what R4 says and deals with supervisory skills. The mechanical operations supervisor is in a project structure. We're rather inclined to take issue with this recommendation.

– The department head: Although someone does it full time, are you saying there's no specificity?

– EDF: No, I'm saying there is. It's managed differently, but it's integrated in the management of a department manager. Your recommendation focuses on the temporary activity.

– The department head: If we rephrase, you're saying that you cover this case. We need proof. Show us that your system is adequately robust.

– EDF: We can cumulate certain discussions. We said in R3 that the occupation is new. While discussing R4, you saw that we're integrating all the tools. We can full well fill out R3 by mentioning the tools.

– The department head: But there's also the aspect of the intermittent activity. If it ties up with R3, then OK. We'll put R5 on the back burner!

Draft recommendation R6 dealt with the methods of determining map targets. During the preliminary preparatory meeting, EDF's representatives had not strongly objected to the experts' analysis. The IRSN coordinator, who did not particularly want to "fight for" for this recommendation, was relatively surprised by the little exchange of views after the reading.

The subject then shifted to skills acquisition and retention methods, for which there were five draft recommendations. The first recommendation concerned the potential effects on safety of training course cancellations. It met with strong disagreement from all of EDF's representatives

> **R7.** The advisory committee recommends that EDF present a report on training course cancellations over the past three years and their potential impact on skills important to safety and radiological protection. This report shall present the means used to offset the effect of such cancellations.

– The department head: This issue falls under the category of further review. It's not within our purview.

– EDF: Is there a point to focusing on this? What's the purpose? There's no problem with the cancellation of training courses. What makes you say that there's a problem with it?

– A representative of the ASN: Training course cancellations are something that can be looked at during inspections.

– The department head: We need to see how these cancellations are done and to what extent. We want to be able to review it.

– EDF: As you said, the matter hasn't been reviewed. But your recommendation is expensive and will create a lot of work for us. What objective information do you have that gives you doubts about it? Is it a major issue that emerged from your review?

– The department head: We'll rephrase the recommendation and see what you think.

Draft recommendation R8 on the recruitment of training staff was introduced as a "further review" by the IRSN coordinator. Since EDF had the information requested, the matter was settled.

The third draft recommendation concerned a matter that had thus far not been brought up. However, the human factors experts had considered it relatively important and thought that it would be a good idea if it were discussed by the members of the advisory committee.

> **R9.** The advisory committee recommends that EDF present, within one year's time, the means adopted to retain rare skills as *baby–boom* employees leave the workplace. EDF shall specify how these means will be disseminated throughout the fleet.

– EDF: You duly noted that we had a skills renewal and knowledge transfer project. We can make an appointment to present the progress of this project to you. That said, this matter is outside the scope of a recommendation! Your recommendation gives the impression that we haven't identified the problem or developed an action plan.

– A representative of the ASN: We want to know what's under way and if it's on schedule. Progress reports are often vague.

– The department head: It stays in the recommendation.

Nevertheless, after the meeting, EDF's representatives proposed a stand/action that was accepted by the experts.

The meeting then turned to draft recommendation R10, which had been brought up on several occasions during the previous meetings. This recommendation did not concern just EDF's representatives.

> **R10.** The advisory committee recommends that EDF bring the actions for determining the relevance of the routine, standard nature of certain initial training or refresher training courses into general use such that skills needs are more adequately met.

– EDF: You saw that there is a nationally approved list of training courses.

– The RSD: The recommendation applies not just to EDF ....

– EDF: No, it does not! Your offensive approach bothers me a little. These are sensitive issues. A training course can't be cut just like that. The phrasing bothers me.

– A representative of the ASN: What is it that really bothers you? That there's too much mandatory training?

– The department head: Exactly. The aim isn't to say that there is too much training. We'll rephrase the recommendation.

Draft recommendation R11, which dealt with the mechanical operations supervisor, did not raise any objections from EDF's representatives.

The following two draft recommendations echoed draft recommendation R4, which had been "rejected" by EDF. But this time both parties seemed to see eye to eye.

> **R12.** The advisory committee recommends that EDF strengthen the measures that support managerial evaluation of activities in on-the-job situations.

– The department head: I have in mind working on the ability to observe and raise the awareness of managers about key points to be looked at.

– EDF: We know for a fact that our situation regarding all the occupations is not satisfactory. But we don't want guys with notebooks in their hands. We would like to stick with managers who aren't overly instrumented.

– The department head: We agree.

– EDF: Going into the field is one of the focal points of our NGD project. It shouldn't become a chore for the managers or employees. We're opting for a stand/action.

The draft recommendation on clearance renewal was the last to be discussed. To no surprise, it elicited reactions not only from EDF's representatives, but also from the ASN's representatives as well as one member of the advisory committee.

> **R14.** The advisory committee recommends that EDF examine the appropriateness of developing possibilities for issuing and partial and temporary revocation of clearance in consideration of the risks incurred by increased transparency of actual practices.

– The IRSN coordinator: We know that lots of actions are possible prior to revocation. Partial or temporary revocation can be envisaged.

– EDF: Clearance is revoked on a local basis. It happens whenever a person does not meet the basic requirements of their occupation.

– The department head: The feeling I get is that there are adjustments and arrangements. Wouldn't it be worthwhile to add safeties to clearance revocation? Regarding the ASN's question, it can't be said that there are no problems with clearance. Because there are.

– An expert on the advisory committee: Couldn't the IRSN look into it more in depth?

– The department head: I think that would be difficult given our external position.

– A representative of the ASN: It's an important matter because it's the last barrier.

– EDF: Placing everything on clearance will create an excessively complex system. We've seen that there are countermeasures. Do you think that they're good? That they're sufficient? I think that you're rigidifying the system.

– A representative of the ASN: Does all this happen behind the scenes?

– EDF: Actions and sanctions are listed in incident reports.

– A representative of the ASN: Are they routinely listed?

– EDF: That's what we're aiming at.

– The department head: We could suggest "that EDF demonstrate that its current system makes it possible to cope with situations of partial or temporary loss of skills."

– A human factors expert: To me, it seems important to stress that clearance is there because the ASN is there. EDF has to make something visible to the inspector. It must be pointed out that, through clearance, the ASN does not inspect everything it thinks it inspects.

– EDF: But what you're saying there is fundamental. You have to have information in order to say that. How can you have an opinion on such serious things!

– A representative of the ASN: Clearance is used to make things seen. If things are done, without revoking clearance, is there a record? Are they visible? If there's a problem, will I have something in writing?

– EDF: I'd like to refocus the discussion a little and put the proposed recommendation back in the historical context. We've always thought that a skills management system was better than a licensing system.

– A representative of the ASN: My problem is do I have the means to know whether a person in charge should in fact not be.

– EDF: The managers ask themselves exactly the same question. They consult the training staff to find out.

– The department head: At the very least, you'll have to explain things. The information in the technical file we have is insufficient. So the recommendation is to explain.

– EDF: We told you already. It's handled. You wrote it. We're not going to write it again!

– The department head: A technical review was not conducted on this aspect.

– EDF: We hadn't realized that you had so little confidence in our system ....

– A representative of the ASN: I'll rephrase it. Either what the system doesn't cover is residual in terms of risks and number, or it's not residual and the question is "what are the arrangements for dealing with it?"

– EDF: OK. There shouldn't be any dramatization effects because we consider that it's residual.

– The department head: We'll rephrase the recommendation. The members of the advisory committee will see that it's a sensitive topic.

The members of the advisory committee would probably have few matters to discuss, for the representatives of the NGD accepted most of the draft recommendations and the contentious issues, or those that the ASN and the experts considered could not be ignored, were few in number. The coordinator was pleased with the meeting's outcome. "The preparatory meeting went along well. The representatives of the advisory committee and the ASN were pleased." (interview of February 15, 2006)

3.5. Conclusion: Nature and Outcomes of the Interactions

All in all, the major changes were the result of the interactions between the experts and the licensees. From the first account by the experts to the NGD's occupation coordinators onward, the process of finalizing the argumentation and explaining the analysis conclusions appeared to be one of agreement and of alignment of the expert and the licensee's viewpoints.

Of course, the day after the preparatory meeting there were still some points argued for by the experts and disputed by the NGD's representatives. Nevertheless, there is no doubt that the various milestones, not least the preliminary preparatory meeting proposed by the department head, made it possible for the parties to see eye to eye. And, as the saying goes, even though people may not agree on everything, they can agree on points of agreement and disagreement. The aforementioned process is illustrated in Table 8. The organization of the columns is chronological.

Statements made and/or problems identified by the IRSN's experts.	EDF's reactions (preliminary preparatory meeting)	The expert's reactions	EDF's reactions (preparatory meeting)	Agreed outcome
Incident selection	Against	None	–	–
Definition of skill	Against	Rephrasing	–	–
Updating and streamlining of the frameworks	In line	–	–	Stand/action proposed by EDF
Uniformization of the maps	Discussion	Slight rephrasing	In line	Stand/action proposed by EDF
Presence of the managers and equipping of their observations	Against	Slight rephrasing	In line	Stand/action proposed by EDF
Access to training and self–expression of needs	Against	Further review, rephrasing of access to mandatory refresher training	Discussion	Draft recommendation upheld
Sustainability of training staff	Against	Further review, rephrasing, alignment with EDF's stance	–	–
Management of supervisory skills	Against	Rephrasing (breaking-down of the theme)	In line	Stand/action proposed by EDF
Clearance renewal	Against	Rephrasing	Discussion	Draft recommendation upheld
–	–	Training postponement report[182]	Against	Draft recommendation upheld
–	–	Methods of dealing with the *senior boom*[183]	Discussion	Draft recommendation upheld[184]

Table 8: Changes in the IRSN and EDF's positions on the issues identified.

4. THE TRANSMISSION STAGE (FEBRUARY–MARCH 2006)

As agreed, EDF's operational coordinator sent his stand/action proposals a few days after the preparatory meeting. The human factors experts then prepared the advisory committee meeting during which the draft recommendations would be discussed. At the end of the meeting, the advisory

---

[182] This issue appeared after the preliminary preparatory meeting.

[183] This issue appeared after the preliminary preparatory meeting.

[184] As we said, EDF's representatives nevertheless proposed a stand/action.

committee's opinion would be communicated to the ASN, which would then address a follow-up letter to EDF.

4.1. The Lead-Up to the Advisory Committee Meeting

The discussions the coordinator had with his colleagues and supervisors qualified his initial, and rather favorable, reactions to the reading of EDF's stand/action proposals. "We held a half-hour meeting with [the assistant director] on February 24. We threw out a few stands/actions. The one on the listing of skills doesn't suit us because they offer nothing new; they just list what they already do around the country. So we're sticking with the recommendation. We wanted them to boost the support, coordination and monitoring of the systems implemented on the sites to develop and maintain mechanical operations supervisors' skills. They list an entire array of things that they already do. We don't see how that boosts anything. We're sticking with the recommendation." (interview of March 2, 2006)

The few days left before the meeting were spent preparing the transparencies that would be presented. The focus to be placed on the occupation studies was an important consideration. "Initially, I didn't want any illustrations of occupations so as not to raise the issue of skills and stay focused on the skills management process. But [the department head] suggested that we show more of what we've got. The occupation of mechanical operations supervisor, for which little has been done, shows negative aspects. So we're also going to talk about the occupations of test technician and control operator." (the coordinator, interview of March 2, 2006)

4.2. The Advisory Committee Meeting of March 14

The five draft recommendations for which EDF's representatives and the IRSN's experts could not (or chose not to) agree on were discussed during the meeting. But before that, the members of the advisory committee discussed the relevance of the scope of the assessment.

Debate on the scope

It may be surprising that the scope of the assessment was the subject of debate on the day of the presentation before the advisory committee. One expert addressed the coordinator sharply. "I understand that it's convenient to look at the skills management system, which is well-defined and has procedures. But wasn't it possible to look more directly at the skills?"

The fear expressed by the coordinator and the department head on several occasions since the start of the review was therefore justified. A representative of the ASN stepped in to voice his support of the choice. "We didn't ask the

IRSN to evaluate the skills. The aim is to find out if EDF has the means to do so itself." Another expert considered that the approach was "perfectly reasonable" given that "the authority can only see over the licensee's shoulder. Furthermore, looking at the tool makes it possible to ensure that things are put into practice, not just theory."

The two social science experts invited to the meeting later criticized the overly instrumental view of the skill. One called it a "push-button view" in which the skills were seen as a stock of materials that should be managed, saying, "We sociologists speak of distributed skills." One expert deplored that the matter was mentioned in only one paragraph in the report, saying that "distributed skills are the invisible mass that binds together everything that is structured. I feel that that's the company's foundation." The head of the SEFH cautioned the members of the advisory committee. "We've conducted many interviews and made many observations but I don't think that an IRSN review can see all that. That's one of its limitations."

The mechanical operations supervisors

The two draft recommendations on the management of the skills of the occupation of mechanical operations supervisor – "the most important of the day's agenda" in the words of one expert – were the subject of a half-hour discussion between the members of the advisory committee.

> **Recommendation No. 1**: Considering the strategic role played by the mechanical operations supervisor, the advisory committee recommends that EDF set up local systems listing supervisory skills so that they may be managed in the departments in the same manner as the other skills.
>
> **Recommendation No. 4**: The advisory committee recommends that EDF reinforce the support, coordination and monitoring of the deployment, on the sites, of the systems for developing and maintaining the skills of the mechanical operations supervisors.

During the discussions, one expert considered that the IRSN's draft recommendation was also in EDF's stand/action. "It seems to me that EDF's stand/action fulfills recommendation R4. EDF was wrong to add the last two paragraphs!"

"I've been hearing EDF's stand/action since 1994," said another expert. "What's needed is something completely new and worded differently because, so far, it has resulted in nothing but small improvements. What I've just said is intended to help the management."

There was concern about the shift prompted by this opinion. "I didn't understand the process at EDF. What is the subject here? Is it recommendations R1 and R4 or things not from the review?"

Whereupon the head of the SEFH said, "We would like to make a recommendation on the skills management process. We noticed during the review that EDF had changed its stance. It is important that the advisory committee back us up."

The advisory committee chairman settled the matter, saying, "I, too, think that we shouldn't put off what can be done today."

An expert rephrased the two drafts as a single recommendation.

> "The advisory committee recommends that, as part of the strengthening of the strategic coordination, central support and tactical monitoring of the deployment, on the sites, of the of the systems for developing and maintaining the skills of the mechanical operations supervisors, EDF set up local systems listing supervisory skills so that they may be managed in the departments in the same manner as the other skills."

"EDF agrees with this proposed recommendation," replied an EDF representative

Postponed training

> **Recommendation No. 2**: "The IRSN recommends that EDF submit a qualitative and quantitative report on the causes of postponements of training in skills important to safety and radiological protection, such as radiological protection training, training of mechanical operations supervisors, physics test training. This report shall cover the past three years and present the means used to offset the effect of such postponements."

An EDF representative spoke up. "You saw a number of absences and objective causes. We cannot reach perfection, but we can say what our areas for improvement are. I don't think that the report you request will lead to any progress. We would rather focus more on the future and see what countermeasures need to be implemented .... Our training program is productive. More than 21,000 employees have been trained."

This prompted a response from one of the experts on the advisory committee. "I find EDF's answer very poor. In the nuclear industry, the figures released are not good. Furthermore, I'm not convinced by the statistical approach. As for the recommendation, the postponement report is just a tool and a tool can be anything. It's managing the effect of these postponements that's important." Another expert expressed his viewpoint. "I don't really see the utility of the postponement report either. What good will it be to know that, for example, 10% of courses were canceled because the trainer had the flu." A third expert chimed in, saying, "I agree that there isn't much point to the postponement report."

The head of the SEFH spoke, saying, "I'm surprised that you let yourselves be won over by what EDF said." To which an expert replied, "We haven't said that we were. I don't give a whit about the causes. What could contribute something is an assessment of the consequences."

Just when it looked as if the draft recommendation was going to die, the EDF representative made a proposal. "Perhaps a little clarification is necessary. We could propose making an assessment for 2006 and seeing how management implements countermeasures to deal with a relatively rare phenomenon (a little more than two percent). We propose that as an interim conclusion to the discussion."

The chairman rapidly scanned the participants and then gave his decision. "Okay. Write it up."

One of the experts spoke up again. "It'll be difficult to do an assessment of the consequences without mentioning the causes. Let's leave it at 'assessment'."

To which an EDF representative said, "I propose using the terms 'countermeasure' and 'arrangement' at the end of the recommendation. Neither means exactly the same thing. To me, arrangement is more organizational." After hearing the opinion of one of the expert's the chairman added his own touch. "That would become systems."

In the end, the recommendation was changed very little. The word "causes" was deleted, and "methods" was replaced by "arrangements".

Self-expression of training needs

> **Recommendation No. 3**: "The IRSN notes that, for some occupations, mandatory refresher training takes up an extensive amount of individual training time, which may lead to inhibition of self-expression of needs. The IRSN recommends that EDF investigate the relevance of the routine, standard nature of certain initial training or refresher training courses such that individual skills needs are more adequately met."

An EDF representative spoke up. "I would like to cite a few figures because there is this impression that people are overwhelmed with mandatory training. Control operators have 12 days of mandatory training a year. This doesn't mean that we're ignoring the issue. I just wanted to get the discussions back on track." The research officer qualified this statement. You're referring to a national requirement. Cite the other requirements! The operators see something different!"

This prompted a discussion between two experts:

– Expert A: I like the concept of this recommendation. That said, I propose removing EDF from it for the reason that it also applies to the ASN and the IRSN.

– An expert B: I disagree with the proposed modification. It's not the advisory committee's job to make recommendations to the ASN even though it obviously should clean up its own back yard.

There followed a lengthy discussion on the use of the terms "mandatory", "routine" and "standard".

– Expert C: Something has to be mandatory.

– Expert B: I have somewhat the same concern: something that is mandatory must be well defined, and thus routine and standard.

– Expert A: I suggest removing "mandatory" from the first sentence and "standard" from the second.

– Expert B: I disagree with both suggestions.

Some were disheartened by the lengthy discussions between the experts. "We're walking a little on eggshells here. I'm not totally convinced that there's a possibility of reaching a real consensus on the matter. There's a risk of pushing things much further than some think. We can stick with EDF's proposal." (an expert)

"If the advisory committee isn't able to come to an agreement, I don't know what we're doing here!" said another expert. "I noted earlier that it was a matter for which EDF requested a recommendation and didn't want to take a stand/action. Let's assume our responsibilities and discuss the wording if we don't agree!" The expert proposed replacing "improve" with "optimize" and removing the word "routine". His proposal was accepted by the chairman and the participants.

Clearances

"I really like the IRSN's analysis but the recommendation leaves me unsatisfied," said one expert. "I'd like for consideration to be given to clearance. Managing skills is good, but managing them under the table isn't. If the public opinion gets wind of a problem, you'll wind up in court! It's unfortunate that there are so few suspensions." This was echoed by another expert. "I agree with what's just been said. The IRSN either goes too far or not far enough. I thought about asking EDF to explain the methods."

This prompted a discussion among several members of the advisory committee:

– Expert A: When there's a problem – and this applies particularly to control operators – clearance is revoked *de facto* because the person is no longer allowed to do their job. But it's not *de jure*. Care must be taken to ensure that the *de facto* aspect is indeed enforced. Then, how is the shift made from *de facto* to *de jure*?

– Expert B: Qualifications are used in manufacturing and are quite clear in the legislation. Revoking a welder's or an inspector's qualification is common. What prevents it from being done in this case?

– Expert C: There are areas of study. There must be temporary restrictions. Legal and administrative research can be conducted. The *de facto* and *de jure* aspects need to be brought together but without canceling out the *de facto* aspect.

"The recommendation formulated as such leads to a discussion with the ASN," said an EDF representative.

An expert: "If I understand correctly, the recommendation suits the licensee pretty well, but that clearance management traceability should be added. Couldn't we just tack on "and management of the clearances in question" at the end? That would open the door to all the possibilities of conditional suspension, etc. But the ball is simply in EDF's court. It might have to show a little creativity or set out how it could temporarily or unconditionally suspend clearance, but it would be traceable."

The EDF representative said that he agreed with the proposed recommendation. With this, the meeting was adjourned. However, the IRSN's rapporteurs, the members of the advisory committee and EDF's representatives remained at the table to draft the advisory committee's draft opinion.

The ASN's follow-up letter was not sent to EDF until September. It contained the positive conclusion stated by the human factors experts and which was echoed by the advisory committee in its opinion. The draft recommendations of the human factors experts – which, modulo the aforementioned changes, had become the recommendations of the advisory committee – were placed in an appendix. They now had the status of requests. EDF's stands/actions were placed in another appendix.

4.3. The End of the Assessment

The final weeks before the unresolved or essential points of the assessment would be presented before the advisory committee were trying for the coordinator. The team of experts and the assistant director held discussions up to the last minute to decide whether to accept EDF's proposals. In a way, their choices were the right ones; all the draft recommendations had become advisory committee recommendations. Like his supervisors and the ASN's representatives, he was very pleased with the course of the assessment. He shared his thoughts during a department meeting the following day. "This is the second advisory committee meeting conducted by the SEFH. Recommendations were discussed. The SEFH is able to conduct advisory committee meetings like the rest."

His satisfaction was shared by all the experts, for the SEFH had done a huge job. Human factors were well and truly legitimate and deserved the same respect given to the engineering sciences. Nonetheless, many human factors experts shared the same strange impression. All – or nearly all – brought up the oddly consensual nature of the meeting ("We couldn't figure out who was representing EDF and who was representing the IRSN.")!. It is true that EDF's representatives agreed with every draft recommendation. One of the experts spoke vehemently, "There was collusion and a blurring of lines. You don't hold an advisory committee meeting if everyone agrees on everything!"

The department head defended himself. "There are two answers to your remark. It's true that, during the scope definition stage, we took out questions liable to cause conflicts. But it must also be said that the analysis is what brought about the consensus. No agreement was reached until the preparatory meeting. They disagreed with lots of things during the preliminary preparatory meeting in December. The quality of the assessment is what won them over. Just because people aren't arguing doesn't mean that safety can't be improved. Conversely, arguing doesn't necessarily improve safety. The reason why fights occur is often because a review went poorly."

The expert retorted, "But a recommendation doesn't mean that it hadn't been dealt with during the review."

The coordinator put an end to the discussion. To him, the consensus between EDF and the IRSN was due primarily to the licensee's level. "The fact of the matter is that EDF is really good when it comes to skills management." (meeting of March 15, 2006)

5. PROVISIONAL SUMMARY

5.1. The Basic Operations of the Human Factors Assessment

The management of the skills of operating personnel in EDF's nuclear power plants was an extension of an issue that had long been known and assessed by the human factors experts, namely training.

Since the aim of the SEFH's assessment, expressed notably by the reactor safety director, was to collect a large amount of empirical data, a team of experts was formed to handle the assessment. The assessment began with a preliminary analysis by the coordinator. Operational incidents characterized by a lack of skills revealed through analyses conducted by EDF or the IRSN were selected by the SEFH's technical adviser to prove the link between skills and safety. A good skills management system was therefore necessary and the experts would have to evaluate the quality of EDF's skills management system. Using a literature review, they established a standard process of skills management of several stages. The literature also emphasized other issues, such as the collective nature of skills and the relationships between human resources policies and skills management. To collect their data, the experts focused on a number of occupations involved in the operation of a plant. They observed the activities of these occupations and conducted "360-degree interviews".

They presented the preliminary analysis to the licensee, whose reactions had two effects on the future course of the assessment. Firstly, it was able to have some themes (HR policies and wage aspects) that appeared to be its preserve excluded from the assessment. Secondly, its apparent interest in the potential outcomes of the assessment led it to "assist" the experts, focus attention on certain occupations and increase the number of sites where field data could be collected.

The hundred-odd interviews and forty-odd observations revealed practices that were more or less unexpected and made it possible to identify problems. The data collected by the experts can grouped be into the following two categories:

– *Flaws in skills management tools and procedures.* Examples: lack of formalization of the methods for acquiring non-technical skills (test technician); lack of traceability of routine training actions (control operator); absence of skills identification and evaluation tools (mechanical operations supervisor). Overall, different targets were used to create the maps and the skills registry was neither uniform nor streamlined.

– *Problems in the skills management process.* Examples: the methods of issuing and renewing clearances (they are not what one thought they were); the surfeit of mandatory training and the difficulty for staff to express their needs (control operator); the difficult task of monitoring; the difficulty in implementing skills evaluation methods.

Relations with the licensee were often strained during the review stage (distrust of the expert's observations, dissatisfaction with instructions often viewed as incomplete). The licensee forced the coordinator to continually "beef up" his argumentation ("You have to separate feelings from facts!"; "The review needs to be supplemented") and regularly cautioned him against deviating from the scope of the assessment determined during the definition stage. In order to stay on subject and answer the questions in the request, the coordinator organized the widespread application.

The subsequent versions of his report took into account the comments and critiques of not only the IRSN's proofreaders, but also of the licensee, who disputed the argumentations and draft requests during two preparatory meetings. This process culminated in a convergence toward common viewpoints, as exemplified by the large number of draft recommendations that became stands/actions. Only one draft recommendation seemed to really pose a problem before the session of the advisory committee – the report on training course cancellations ("Your recommendation is expensive and will involve a lot of work for us"; "There are no problems with this"). And yet, despite experts on the advisory committee who seemed hostile to it ("this assessment is meaningless"), the draft recommendation was "rescued" by a representative of EDF (albeit by cutting the analysis period from three years to one). The SEFH experts who had not participated in the assessment were surprised by the session's consensual outcome.

5.2. Comparison of the Empirical Data and the Theoretical Models

Unsurprisingly, this third account leads us to question the canonical model. Like the other cases, the case at hand does not fit the canonical model for a number of reasons: its collective nature, the limited freedom of choice of its operating protocols, the range of knowledge used by the expert and which was built during the assessment. When the coordinator drew a distinction between the research officers from the SEFH and the external research officers, he realized that the assessment required expertise in the assessment's institutional context and experience in practicing a mechanism in order to be able to understand its subtleties. This is yet another illustration the non-independence of the learning of the human factors expert and the assessment process.

As in the case of the safety review of the Minotaure reactor, the assessment was a planned process, punctuated by many meetings (kick-off, internal consultation, scope definition, feedback to the sites, feedback to the central services, halfway, preliminary preparatory, preparatory, advisory committee) and report proofreading and validation stages. Given the many comparison meetings held with the licensee before the assessment conclusions were drawn up, the adversarial procedure was remarkably well integrated in procedures. However, the values of transparency and independence were nowhere to be found. One can wonder if the composition of the advisory committee had anything to do with this, for some of its members were active EDF executives. It is also worth mentioning that EDF's representatives present at the advisory committee meeting participated in drafting the opinion of the standing advisory group.

Our observations show that there were two types of control: for the minor part, result-based control, as evidenced by references to past incidents; and, for the major part, procedure-based control, as evidenced by the use of a standard process that enabled the experts to identify the "flaws" in EDF's skills management system.

As previously, the assessment's operations were not limited to review. The investigation of the operation of the departments from the perspective of skills management is not directly similar to the review. Through their investigations, the experts were able to bring to light practices that were not in the specifications for the skills management tools. In doing so, they created a source of learning for both the experts and the licensee.

Perhaps more than in the other two cases, the issue of whether the expert was captured is worth raising. In a certain way it was when, the day after the advisory committee meeting, one of the SEFH experts criticized the behavior of the IRSN's experts and the licensee's representatives ("collusion, blurring of lines!"). Indeed, not only did EDF make the strange move of rescuing a draft recommendation, but on several occasions the coordinator served as the licensee's mouthpiece and advocate by answering questions or criticisms leveled at EDF.

As we saw in the case of the Minotaure review, the licensee curbed the expert's freedom of action on several occasions. For example, during the scope definition stage it was able to considerably limit the assessment's scope. Over the course of the review it routinely attempted to control the work of the experts, who did not always let themselves be told what to do (such as with the dissemination of the occupation study reports). They sometimes had the unpleasant feeling of being regarded as internal auditors. They made this known and, on several occasions, displayed their independence by not conforming to the wishes of the EDF coordinator, such as when he wanted a number of passages expunged from the report (definition of skill and references to certain incidents).

Nonetheless, it may be wondered if adversarial procedure was really at work. The licensee's involvement in the writing of the assessment's conclusions is startling, if not unsettling. During the preparatory meeting, whenever the licensee agreed with one of the expert's draft requests, it rephrased it and the expert had to give his opinion on the stand/action. Any draft request that the licensee disagreed with became a draft recommendation to be discussed during the advisory committee meeting. The licensee stated at the end of the discussion that it agreed with the recommendation. A final example occurred during the preparatory meeting. The licensee more or less agreed with a draft recommendation and wanted to draft a stand/action. However, the expert preferred that the matter be discussed by the members of the advisory committee. It was thus presented as a proposed recommendation. This method of operation gives the impression that both parties were looking for a compromise. And yet the many examples of the expert's independence of judgment make it difficult to assert the occurrence of capture.

Conclusion to Part Two: The Singular Aspects of the Assessment Factory

When compared with classic theories of assessment and control, the observation of the human factors experts at work on three cases representative of their activity, produced novel findings.

Indeed, it is clear that the human factors assessment does not fit any of the scientific assessment models in the literature. Although the canonical model was quickly abandoned, neither do our data indicate the characteristics of the procedural model despite what the inventory of the many procedures in the IRSN assessment could have suggested. Only the adversarial procedure was embodied in procedures and in just two cases. Furthermore, since it is confined to technical dialogue institutions, the assessment studied is also far removed from the "hybrid forum" model predicated on the integration of many different types of knowledge[185].

We had sensed that the experts would use many forms of control during the review stage. Our observations revealed the use of result-based control (use of OEF data) and procedure-based control (human and organizational barriers assumed to ensure safety). We also identified traces of clan control in the first account (evaluation of the human factors skills implied by the licensee). Our data also show that the assessment cannot be boiled down to a sequence of inspection operations. In the first two cases, the analysis was focused on understanding the event chains leading to the incident, while the third was focused on understanding the organizational workings.

Our answer to the question of whether the expert was captured by the licensee is 'no', albeit with a few reservations. Firstly, we were present while the experts formed and wrote their opinions without any interference from the

---

[185] cf. Callon, M. and A. Rip (1991). *Forums hybrides et négociations des normes socio-techniques dans le domaine de l'environnement. La fin des experts et l'irrésistible ascencion de l'expertise.* Environnement, science et politique. Les experts sont formels, Paris, Germes.

licensee. Secondly, the experts often stood firm during negotiations with the licensee. However, the licensee's interference in the choice of objects and methods and the wording of the assessment's conclusions – such as in the case of the assessment of EDF's facilities – shows that the experts' independence was routinely jeopardized during the assessment. In the case of the third assessment, it depended greatly on the determination of the experts to not let themselves be captured and become the licensee's contractors.

This last point illustrates an aspect of the human factors assessment that is not shown by the three cases studied – namely the impact of the human variable on the course of these assessment processes, or what we referred to as the "expert dependence" of the human factors assessment. The value of our accounts – which show how much the invocation of a system of procedures and the listing of a set of recommendations were inadequate to explain the assessment – is to bring to light a certain complexity of the production system due in particular to the subtleties of the relationships between the various parties involved (collaboration, trust, negotiation, compromise, conviction, agreement, failure to understand) that gives many operations in the process their novelty. It is these singularities, which continuously punctuate the production system, that make it unpredictable[186]. In a way, the production system is its own product, with some of its stages producing others that are not necessarily planned. Lastly, by revealing the many instances of the technical dialogue between the experts and the licensees, these singular aspects show a control method inherited from "French cooking".

Nevertheless, do the historical and institutional influences suffice to justify this exploratory process made up of unstable relationships? We think not. By following Armand Hatchuel (2000), who considers that the two components of any collective action – knowledge and relationship – are inseparable, we believe it is necessary to take a closer look at the knowledge and relationships on which the human factors assessment is based. Would the low level of knowledge explain the singular nature of the relationships at play?

---

[186] During one meeting of the research steering committee, a participant made an analogy with Brownian motion, a stochastic process used to simulate the random, continuous movement of very fine pollen grains suspended in the air as well as changes in the prices of financial securities.

Part Three.
The Effectiveness of Assessment

Our account of how human factors assessments are produced identified and described the many interactions and operations that culminate in a set of recommendations. In this third part, we will look at these recommendations more closely in order to answer three questions: 1) What knowledge are they based on? 2) What are their effects? and 3) Should these recommendations be viewed as the sole outcome of the assessment process? To answer these questions we will propose three types of assessment effectiveness, i.e. rhetorical, cognitive and operational.

Listing the types of knowledge used during the review makes it possible to see their weaknesses. As a review of the scientific literature shows, this deficiency, which in part explains the specificities of the relationships between experts and licensees, cannot be corrected by references to academic research. Thus, in order to respond to questions, critiques and objections and ensure that their draft recommendations clear each stage (development, data collection, proofreading, validation, transmission) the human factors experts are led to focus on the "rhetorical effectiveness" of the assessment, sometimes to the detriment of exploring new types of knowledge and investigating the links between human and organizational factors and safety, which are sources of cognitive effectiveness. [Chapter 6]

To examine the effectiveness of expert assessment, the impacts of human factors assessment on nuclear facilities must naturally be studied. Together, these effects can be tied to a third dimension of the effectiveness of assessment, referred to as operational, which is far from easy to evaluate. This requires viewing the assessment as both the element of a series of assessments and a process of interactions between regulators and the regulated, some of which alone produce effects on facilities. As we list the effects on the assessment, we will identify the skills the expert must master in order to produce them and summarize the types of knowledge required for an effective assessment. [Chapter 7]

We will then be able to formulate proposals likely to restore the balance of the effectiveness of the human factors assessment according to the three aforementioned dimensions. [Conclusion to part three]

Chapter 6. Persuade or Convince: the Rhetorical and Cognitive Effectiveness of the Assessment

The look at the history of the Department for the Study of Human Factors (SEFH) made it possible in particular to see the difficulties encountered by the project that warranted the creation of laboratory at the institute, namely "to obtain a fuller understanding of the behavior of operators at nuclear facilities (reactors, laboratories, plants, etc.) in order to be able to propose improvements in all areas that may contribute to enhancing safety, such as the display of information in control rooms, written procedures, operation and troubleshooting tools, organization of work, perception of risks and training."[187] Despite this failure, as our examples show, the human factors assessment does indeed exist and address a variety of valid subjects and its recommendations apply to nuclear facilities of all types. However, analyzing the recommendations from the various cases and placing emphasis on both types of analysis – causal analysis and analysis through comparison with a reference organizational model – will lead to underscoring weakness of the established knowledge. However, blame should not be laid on the human factors experts. Even though the findings of the scientific literature provide clarifications and lead to caveats, they enhance the state of knowledge in the field only marginally. What type of analysis is best used to perform an assessment? It seems that causal analysis, which clarifies the ties between human and organizational factors and safety of nuclear facilities, makes it possible to achieve higher rhetorical effectiveness – probably by getting the stakeholders on the same page, the chief objective of the

---

[187] Gomolinski, M. (1985). Paramètres humains dans la sûreté des installations nucléaires, CEA/IPSN/DAS/LEFH: 5, p. 1.

assessment – and higher cognitive effectiveness. Nevertheless, several "good reasons" justify analysis through comparison with a reference model[188].

1. THE TYPES OF WEAK KNOWLEDGE IN THE HUMAN FACTORS ASSESSMENT

In order to find out what types of knowledge were used by the human factors experts, we looked at the conclusions and recommendations in their assessments. This data seemed to us more reliable and better suited than information on the experts' careers or training or even their reference works or publications. The analysis will enable us to establish an orderly set of human and organizational safety factors that will form a reference model. Examining the arguments for these safety factors will enable us to show the existence of two different, yet possibly complementary, standard analyses: causal analysis and analysis through comparison with a reference model. The first is exploratory in nature and can make it possible to establish a strong cognitive link between the human or organizational factor and the safety of the facility being assessed. The second, used often by the experts, is predicated on the assumed benefits of the reference model. However, clearly defining this model's components and evaluating its effects on safety seem difficult. It is in this sense that the established knowledge of the human factors assessment is weak.

1.1. *The Human and Organizational Variables Associated with the Recommendations and Conclusions*

What objects did the human factors experts look at in their assessment? Using an analysis of the recommendations expressed in several assessments, we will explain the components of a model of safe organization as viewed by the experts[189] . Most of their recommendations were for improvements that consisted of bringing the organization assessed (a facility or a group of facilities) more in line with this reference model, which places great emphasis on management rules, systems and tools.

---

[188] We will primarily use the data from the previous part to support the arguments in this chapter. We will also refer to the literature on risk management and organizational reliability, classical findings from research in management and organizational theory and Chaïm Perelman's concepts of the new rhetoric (Perelman, C. and L. Olbrechts-Tyteca (1958 (2000)). *Traité de l'argumentation*. Bruxelles, Editions de l'Université de Bruxelles, Perelman, C. (1977 (2002)). *L'empire rhétorique. Rhétorique et argumentation*. Paris, Librairie philosophie J. Vrin).

[189] We will incorporate the various components used by each expert in their assessment.

The analysis of the recommendations and conclusions shows that *human and organizational variables* can be associated with the experts' recommendations. This term is in fact similar to that of *design parameter* used by Henry Mintzberg[190] to develop his theory of organizational structuring. However, "variable" illustrates the prescriptive mechanism better in that the requirement assigns a target value to the variable that is meant to improve safety. Like Mintzberg, we have categorized these variables under several headings[191] in Sidebar 23.

- **Operating experience feedback system**: incident and malfunction analysis methods; integration of human factors; methods for the centralization and sharing of analysis findings.
- **Human-machine interfaces:** interface design methods; integration of human factors; ergonomics of existing interfaces.
- **Management of the documentation system**: operating documentation design methods; quality and consistency of the documentation; set of systems for encouraging and checking its use.
- **Skills management process**: set of systems and tools for managing technical and non-technical skills and clearances; training in risk and safety issues.
- **Organization of work**: all the arrangements for the coordination, transmission and dissemination of information; definition of roles, assignments, responsibilities, delegation methods, inspection methods; workload and working conditions.

Sidebar 23: Five groups of human and organizational variables.

This was done for the assessments described in the second part as well as two other assessments in our sample (Artémis review, operating organization). The analysis looked at the various recommendations in the form of requests, commitments or stands/actions. We also took into consideration the variables that were mentioned during the reviews we followed and were not formulated as explicit recommendations.

The first five recommendations of the Minotaure review assessment bring a human and organizational variable into play. The last two, however, are of a different sort (cf. Table 9) in that the licensee is not asked to explain or change the value it assigned to a human or organizational variable but rather to clarify and analyze the hazardous activities and related risks. The expert also

---

[190] Mintzberg, H. (1978 (2005)). *Structure et dynamique des organisations*. Paris, Editions d'Organisation, Mintzberg, H. (1980). "Structure in 5's: a synthesis of the research on organization design." *Management science* **26**(3): 322–341.

[191] Mintzberg identified a set of nine design parameters grouped under the following four headings: 1) Position design (job specialization, behavior formalization, training and indoctrination); 2) Superstructure design (unit grouping, unit size); 3) Lateral structuring (planning and control systems, liaison devices); 4) Design of decision-making systems (vertical decentralization, horizontal decentralization).

emphasized other variables in his conclusions, particularly the definition of responsibilities and reporting lines (which, according to the licensee, were meant to justify the reorganization), the sharing of information among the crews (which prompted concern from the expert about the reorganization) and the definition and performance of inspections of hazardous activities (particularly during fuel rod fabrication). These three variables can be placed under the heading of "organization of work".

Recommendations	Associated variables	
	Headings	Variables
Improve the consideration of human and organizational factors in the analysis of incidents and anomalies	*Operating experience feedback system*	Formalization and integration of HFs
Establish and provide the method for drawing up the operating documentation and validating its readability	*Management of the documentation system*	Design method
Boost criticality risk training	*Skills management process*	Training
Establish and provide the method for identifying skills needs and training means selected		Method for identifying needs and means
Collect and provide feedback on the performance of the organization recently put into place	*Organization of work*	Reorganization evaluation method
Perform and provide an in-depth analysis of the reactor operation and the rod handling operations	-	
Complete the analysis of human activities sensitive to the criticality risk	-	

Table 9: Recommendations and associated variables (Minotaure review).

In the case of the Artémis incidents assessment, variables other than those listed in Table 10 were mentioned. These were in particular the definition of the response methods used by the RPU's responders and the definition of responsibilities and coordination and communication methods between responders, particularly those on different crews. These variables, which had been addressed by corrective actions expressed by the facility manager, can be classified under the heading of "organization of work".

Recommendations	Associated variables	
	Headings	Variables
Systematically analyze the preparatory stage of the operations identified in the incident reports as having led to the incident	*Operating experience feedback system*	Integration of the preparatory stage
Review the ergonomics of all workstations likely to have an impact on the safety of the facility	*Human-machine interfaces*	Quality of existing workstations
Define arrangements (awareness building, audits, etc.) that ensure that the modification procedure is correctly followed	*Management of the documentation system*	Systems for encouraging and checking use
Provide progress on the drafting of the operating documents		Design method
Broaden the scope of the modification request procedure	*Organization of work*	Preparation and inspection of hazardous activities[192]
Include a risk analysis with every modification request		
Justify the (effective) corrective actions relating to operations managers		Matching of resources to workloads

Table 10: Recommendations and associated variables (Artémis incidents assessment).

Table 11 shows the variables for the Artémis safety review. In addition to them, mention was made during the review stage – in which we participated – of the necessity for efficient coordination and a clear definition of the limits of accountability between the researchers and the operations unit ("coordination methods", "definition of responsibilities" and "assignment of roles" variables). The definition of the researcher evaluation systems was also mentioned. These variables can be grouped under the heading of "organization of work".

---

[192] In a way, these two recommendations are also akin to those in the Minotaure review assessment on in-depth analysis in that the researchers and licensee are expected to explain and prevent risks related to their activities.

Recommendations	Associated variables	
	Headings	Variables
Supplement the document management process with arrangements for identifying the consequences of changes to instructions and for verifying the applicability of a modified instruction	*Management of the documentation system*	Quality and consistency of the operating documentation
Train safety engineers in human factors	*Skills management process*	Training of safety engineers
		Integration of human factors
Develop a researcher training plan (make more extensive use of training through mentoring and include hands-on incident condition management exercises)		Training through mentoring
		Training in incident conditions
Draw up a report on the performance of tasks by operations managers (matching of resources and the necessity, or otherwise, of maintaining the occupation)	*Organization of work*	Definition of the position
		Matching of resources to workloads

Table 11: Recommendations and associated variables (Artémis review).

Among the variables mentioned during the skills management assessment – these variables are not listed in Table 12 and can be grouped under the heading of "skills management process" – are the quality of the system for managing non-technical skills, the matching of "training resources" to needs, and the matching of staff turnover time to the skills acquisition period. The last two variables are, respectively, variants of the "quality of the training system" and "quality of the skills management system" variables.

Like the skills management assessment, the variables "handled" during the operating organization assessment are grouped under a single heading, i.e. "organization of work" (cf. Table 13).

Recommendations	Associated variables	
	Heading	Variables
Set up local systems listing supervisory skills		Listing systems
Streamline national skills management documents		Update and harmonize the documentation
Draw up a qualitative and quantitative report on the postponement of training		Quality of the training system
Investigate the relevance of the routine, standard nature of certain initial training or refresher training courses	*Skills management process*	Amount of mandatory training
Define and formalize the best method components for determining skills targets		Quality of the skills management tools
Present the methods used to compensate for a temporary or partial deficiency in skills, particularly how crews are formed to perform activities sensitive to safety and radiological protection, and management of the relevant clearances		Clearance management process

Table 12: Recommendations and associated variables (skills management assessment).

Recommendations	Associated variables	
	Heading	Variables
Improve the safety engineer's direct access to information so that he can independently conduct analyses		Delegation system
Draw up a report on the operations manager's performance of his entire range of duties (ability to fulfill his duties as required, ability to know when to assume his safety duties and when to delegate them)	*Organization of work*	Delegation system Matching of resources to workloads
Evaluate the impact of the new distribution of verification duties during an outage on the safety engineers' workload		Matching of resources to workloads

Table 13: Recommendations and associated variables (operating organization assessment).

Each of the human and organizational variables found were assigned the target value identified by the experts in their recommendations and then grouped together in Table 14[193]. The values make these variables human and organizational factors of nuclear safety as conceived and recommended by the experts.

Operating experience feedback system	– "well-crafted" operating experience feedback that incorporates human factors – the preparatory stage is analyzed in incident reports
Human-machine interfaces	– an interface design method (formal specifications, ergonomic validation) that takes human factors into account – first-rate interfaces
Management of the documentation system	– a "good" document design method (that incorporates in particular a readability validation stage) – a coherent documentation system – formal, quality procedures that incorporate a risk analysis stage – known and followed procedures
Skills management process	– "good" safety training for operators – human factors training for safety engineers – a "good" quality training system (training through mentoring, training in incident conditions, mandatory amount of suitable training, sufficient resources) – "good" management of technical and non-technical skills (quality, consistency, exhaustiveness of the systems)
Organization of work	– "clearly" defined duties – a "well"-defined distribution of roles – "well"-defined responsibilities and reporting lines – a suitable delegation system – suitable methods of coordination, particularly among the various crews – "good" dissemination of information – "effective" inspection methods – individual assessment systems that are in accordance with safety requirements – "good" working conditions (resources matched to workloads, high-level occupations and duties) – a "good" organizational design method

Table 14: The reference model.

This reference model in particular calls for the following comments:

- *Generally speaking, the human factors experts looked at the organization from a technological standpoint, that is, by placing emphasis on the rules, systems, management tools (e.g. design method, skills management system, delegation system);*

---

[193] Although the proposed list, which is based on the analysis of five cases, may not be exhaustive, interviews of several human factors experts indicate that it is adequately representative.

- *The imprecise nature of the target values of several variables of the model is worth emphasizing (e.g. "good" dissemination of information);*

- *This reference model is not set in stone. This was seen in particular in the case of skills management, where some elements were made up during the scope definition stage. The SEFH's experts have amassed these elements and now use them for other assessments (such as for safety reviews);*

- *These human and organizational factors are part of a story of nuclear safety thinking. Indeed, they can be associated with various levels of defense in depth, with the observation that they relate more to prevention than the mitigation of consequences. Furthermore, the significance of the organizational and managerial themes is noteworthy. It illustrates the extension of the assessed objects mentioned in Chapter 2.*

The experts used (a subset of) these human and organizational factors in their recommendations. The analysis of their supporting arguments is also enlightening.

1.2. Justifications of the Recommendations

How did the experts justify their recommendations, requests for justifying or changing the value of the human or organizational variables? A look at the arguments used in the cases that we followed reveals two types of analysis used by the experts: analysis through comparison with the reference model and causal analysis. We have compiled the main arguments for the recommendations for each of the three cases described in Chapter 2. Our analysis reveals three types of arguments: those that emphasize a gap between the data assessed and the reference model; those that are grounded in an element of an event chain that could, or did, lead to a hazard; and those that express an insufficiency in the data assessed. In the case of the skills management assessment, we also mention the arguments that justify the assessment's very value.

Many of the expert's arguments in the assessment of the Minotaure experimental reactor are of the first type (cf. Table 15). They show a gap between the reference model in Table 14 and data assessed. Remember that the expert had not explained this model: "This model contains the good practices. Some of the findings come from the literature. It is known that reorganization can lead to negative effects and that there are always interface issues among the different crews. Another example is that it is important for a human factors project to be managed just like a normal project and that deliverables are defined. Lastly, although it's good that operating staff participate in the writing of instructions, that isn't enough. An inadequately defined role is a risk factor. None of that is new." (the expert, interview of November 21, 2005)

	Gap with the reference model	Element of a hazardous event chain	Insufficient data
Lack of justification of choices related to reorganization	X		X
Principle of continual improvement	X		
No formalization of the discussion on skills management	X		X
No operating documentation design method with a validation stage with the operators	X		
No involvement by the CEA's human factors experts in the analysis of the failures	X		
No human factors chapter in the OEF file	X		X
No identification and analysis of the incidents' root causes and human errors	X		
Incomplete analysis of the risks of the operations activity in the analysis			X
Possibility of carrying clearances simultaneously for two different positions necessary for fuel rod fabrication	X	X	
Incomplete analysis of the risks of the fuel rod fabrication activity		X	X
Importance of the criticality skills management process	X	X	
Lack of formalization of the process for validating the ergonomics of the control room renovation	X		

Table 15: Categorization of the arguments (Minotaure review assessment).

A few of the expert's recommendations were not justified solely in reference to an organizational model. When giving his opinion on the fuel rod fabrication process, he observed several sequences for which he identified risk scenarios, for which he received the backing of the coordinating generalist engineer and the criticality expert. This led the expert and his supervisor to identify scenarios not cataloged by the licensee and to consider that more demanding inspection methods might prevent incidents from occurring.

Lastly, although one of the notable effects of the proofreading process was to make it shorter, several arguments focused on the quality of the file provided by the licensee. Certain missing information (human factors chapter in the operating experience feedback file) and the poor quality of the documents (operational activities analysis) played a part in justifying several of the expert's recommendations.

	Gap with the reference model	Element of a hazardous event chain	Insufficient data
Poor ergonomic design of a cabinet	X	X	
Response methods used by the RPU's experts	X	X	X
No radiological testing equipment	X	X	X
Severity of the potential consequences of the incidents		X	
No documentation	X	X	
No lineup checks	X	X	
Importance of the risk analysis and preparation	X	X	
Preparation not taken into consideration in the incident analyses			X
Poor quality of the modification procedure		X	
Constraints placed on the operations managers	X		
No justification of the working conditions of the safety engineers and operations managers	X		X

Table 16: Categorization of the arguments (Artémis incidents assessment).

In the case of the Artémis incidents assessment, most of the recommendations were justified by an analysis of the event chains that led to the incidents (cf. Table 16). The expert worked closely with the generalist engineer to conduct this analysis. Some event chain elements identified are also out of line with the reference model (lack of preparation, inspection, for example), which was partially used by the expert and his supervisor during the scope definition and review stages. Although it is appropriate to distinguish two methods of construction of the arguments, the reference model is a heuristic tool used to piece together event chains. Furthermore, the expert told us he had a few uncertainties about his data. In particular, he felt that he did not have enough information on the response methods used by the RPU's experts. He cited this lack of data as reasons for some of his recommendations.

Mention was made of past incidents in the assessment of the evaluation of the management of the skills and clearances of operating personnel at nuclear power plants. The aim in this case was to justify the theme analyzed. "The incidents and significant events that occurred at EDF's PWRs show that some situations sometimes occur rather differently from expected. Regardless of the support provided by instructions, the organization and the human-machine interfaces (HMI), the operators' skills are nevertheless used and are a major factor in the safety of the facilities. The review of the methods of managing skills and assigning clearances thus plays a significant role in safety in that, in short, the safety of the facilities partially resides in the operators' skills." (report excerpt) Several operational incidents that occurred during the operation of the nuclear power plants substantiate this. The following argumentation was adopted: since skills are a "major factor of safety", they must be managed appropriately, and this requires setting up a good skills management process. So the experts established a standard process of skills management after conducting a review of the scientific literature. We stress that the benefits of such a body of formal systems and its effects on skills are still the subject of academic debate[194]. In recent years, particular emphasis has been placed on less formal types of learning[195], collective learning[196] and organizational learning[197].

	Gap with the reference model	Element of a hazardous event chain	Insufficient data
Non-uniformity and large size of the skills registry	X		
Lack of updates to the national frameworks	X		
Different methods of determining the skills target (map)	X		
Poorly equipped manager observations-evaluations	X		
Glutted skills acquisition system (large amount of mandatory training)	X		

---

[194] Education_permanente (2003). "Où en est l'ingénierie de la formation ?: dossier." (157).

[195] Blackler, F. (1995). "Knowledge, knowledge work and organizations: an overview and interpretation." *Organization studies* **16**(6): 1021–1046, Chatzis, K., F. de_Coninck, et al. (1995). "L'accord A. Cap 2000: la "logique compétence" à l'épreuve des faits." *Travail et emploi*(64): 35–47, Nonaka, I. and H. Takeuchi (1995). *The knowledge-creating company: how japanese companies create the dynamics of innovation.* Oxford, Oxford University Press.

[196] Hatchuel, A. (1994b). "Apprentissages collectifs et activités de conception." *Revue française de gestion*(99), Fixari, D., J.-C. Moisdon, et al. (1996). Former pour transformer: la longue marche des actions de requalification. Paris, Centre de gestion scientifique de l'Ecole des mines de Paris: 74.

[197] Argyris, C. and D. A. Schön (1992). *Theory in practice: increasing professional effectiveness.* San Francisco, Jossey-Bass.

Inadequate identification of supervisory skills needs	X		
Uncertainty about the long-term retention of training staff	X		X
Difficulty in obtaining certain types of training	X		
Postponement of certain types of training	X		
Non-presentation of the results of the skills renewal project			X
Near-routine renewal of clearances	X		
Non-auditability of alternative measures to non-renewal of clearance	X		X

Table 17: Categorization of the arguments (skills management assessment).

However, this link with this safety skills management model was not discussed further. Nearly all the recommendations were made in relation to this ideal process (definition and identification of needs, identification of skills maintenance or acquisition needs, investigation of solutions, implementation of maintenance or acquisition actions, evaluations). Most of the experts' arguments revealed gaps with this ideal process and are incorporated in our reference model (cf. Table 14).

Thus, independently of the methods used to collect their data, many arguments justifying the experts' recommendations stemmed from an analysis through comparison with the reference model and/or a causal analysis, i.e. an analysis of the event chains that led, or could lead, to an incident. Furthermore, the analysis can be supplemented with arguments on the quality or quantity of the data collected. Such data can be 1) data not specified by the licensee; 2) data deemed to be of insufficient quality by the expert; or 3) data on a specific issue identified at the end of the review or that is tricky to investigate. In the first two cases, the argument can often be construed as being out of phase with the reference model (e.g. the missing human factors chapter in the Minotaure OEF file and the lack of justification for the long-term retention of training staff for the skills management case). In the third case, it is often an insufficiently investigated component of a causal analysis (e.g. response methods used by the RPU's experts at the Artémis facility).

All (or nearly all) the arguments of an assessment are thus derived from an analysis through comparison with the reference model and a causal analysis. The soundness of the arguments justifying the enhanced safety that implementing the requirement is meant to bring about depends on the analysis used. The concept of cognitive link clarifies this significant difference between the two standard analyses.

1.3. Cognitive Link with Safety and Cognitive Strength of the Arguments

A strong cognitive link is said to be established between a human or organizational factor and the safety of a facility when implementing this factor would have prevented a past incident (or mitigate its consequences) or might prevent a specific potential incident scenario (or mitigate its consequences). When the argument justifying a requirement is predicated on an incident scenario (past or potential), the causal link between this requirement and the safety of the facility is, in a way, demonstrated. Say an unchecked lineup error causes an incident. Improving lineup checks with a set of human and organizational factors (awareness, procedure, double-checks, training, etc.) will decrease the probability of occurrence of a lineup error and thus increase safety. Likewise, effectively improving lineup error checking systems that were found to be insufficient will enhance safety since lineup errors will have little likelihood of resulting in hazards.

On the other hand, the argument revealing a single gap with the reference model raises a few questions. Although it is understandable that implementing human and organizational factors characterizing this model is a step "in the right direction", it must be acknowledged that their causal relationship with safety is less established. For example, one can hardly deny that "the proper dissemination of information enhances safety". But how to go about evaluating this alleged enhancement? Shouldn't the definition of poorly disseminated information be clarified further? And just how could this insufficient dissemination cause an incident? Once can see that it justifying the effects of human and organizational factors on safety is not an easy task. Likewise, justifying a recommendation based on the insufficiency of data does not enable direct link between the recommendation and the safety of the facility to be established. Such arguments are said to be cognitively weak.

We would like to make clear the two analysis types discussed are not exclusive. As we have said, the reference model can guide the causal analysis and the causal analysis can supplement an analysis through comparison with the reference model. However, the causal analysis is necessary to establish a strong cognitive link between an organizational factor and safety.

Our discussion is summarized by the following two assumptions:

- *An argument stemming from a causal analysis is said to be cognitively strong;*

- *An argument stemming from an analysis through comparison with the reference model is said to be cognitively weak.*

1.4. Conclusions: Weaknesses of the Established Knowledge

This distinction between the experts' arguments thus leads us to narrow the justificatory scope of the analysis through comparison with the reference model. Moreover, we have seen that both analysis types led to using the reference model's human and organizational variables. It must be noted that how values are assigned to these variables often remains unclear. Although the effects of "proper dissemination of information" are viewed as positive, the attributes of this organizational factor ought to be made clear. What does proper dissemination of information mean? When made clearer, these attributes often lead to increasing formalization – the standardization of behaviors by means of procedures. And yet it is acknowledged that overformalization can be counterproductive, which is exactly what the reactor safety director emphasized about the evaluation of skills by managers.

This leads us to question the reference model, which was used by the experts regardless of the analysis type. Evaluating the effectiveness of its human and organizational factors on safety remains tricky to establish without a clearer explanation of their attributes and the incident scenarios they may avert. Perhaps the literature provides additional justifications that can answer the following questions: How do researchers establish ties to safety? Do they view the reference model variables as being legitimate? Are there others that should hold the experts' attention?

2. RECOMMENDATION-PROOFED LITERATURE

Industrial safety issues entered the academic arena mainly via the engineering sciences[198] – and particularly one area of operational research: reliability – and ergonomics. Spectacular accidents in the transportation, chemicals, oil and nuclear industries drew the attention of researchers. The 1980s marked the beginning of the development of theories, approaches and tools that attempt to analyze and show the contribution of human and organizational factors to safety. Today, the literature abounds with publications across many disciplines (notably psychology, ergonomics, sociology, management, political science and engineering sciences) on this.

Although the causal analysis is commonly used, especially to reconstruct event chains that led to an industrial accident (*a posteriori* analysis), some

---

[198] Villemeur, A. (1988). *Sûreté de fonctionnement des systèmes industriels.* Paris, Eyrolles, Rolina, G. (2005). Savoirs et discours de l'expert. Le cas du spécialiste des facteurs humains de la sûreté nucléaire. *14e rencontres "Histoire & Gestion".* Toulouse.

researchers have mentioned the limits of this approach[199]. Using an ideal model – which the SEFH's experts do – can circumvent these difficulties. Several publications with a resolutely prescriptive purpose highlight organizational barriers that would avert hazardous events or mitigate their consequences[200]. We will use the famous model developed by the psychologist James Reason[201] to present this research, for we consider its principles, critiques and uses to be emblematic. In order to identify what would help guarantee the safety status of hazardous facilities, other researchers choose to observe daily practices, under normal or slightly abnormal conditions, of operating crews, maintenance crews or design project teams in hazardous facilities[202]. Although they highlight the difficulty of apprehending safety, some have developed heuristic concepts and others have identified characteristics of what they refer to as "high-reliability organizations". Although a conception of safety based on a set of organizational barriers is quite similar to that of the experts, the concepts and methods used

---

[199] Dodier, N. (1994). "Causes et mises en cause. Innovation sociotechnique et jugement moral face aux accidents du travail." *Revue française de sociologie* **35**(2): 251–281, Perrow, C. (1994). "The limits of safety: the enhancement of a theory of accidents." *Journal of contingencies and crisis management* **2**(4): 212–220, Journé, B. (1999). Les organisations complexes à risques: gérer la sûreté par les ressources. Etude de situations de conduite de centrales nucléaires. *Sciences de l'homme et de la société. Spécialité Gestion*. Paris, Ecole Polytechnique. **Thèse de doctorat**: 434.

[200] Reason, J. (1997). *Managing the risks of organizational accidents*. Aldershot, Ashgate Publishing Limited, Plot, E. (2007). *Quelle organisation pour la maîtrise des risques industriels majeurs ? Mécanismes cognitifs et comportements humains*. Paris, L'Harmattan.

[201] Reason, J. (1990). "The age of organizational accident." *Nuclear engineering international*: 18–19, Reason, J. (1993). *L'erreur humaine*. Paris, Presses universitaires de France, Reason, J. (1997). *Managing the risks of organizational accidents*. Aldershot, Ashgate Publishing Limited, Reason, J. (2006). Human factors: a personal perspective. Helsinki.

[202] Roberts, K. (1990). "Some characteristics of one type of high reliability organization." *Organizations science* **1**(2): 160–176, La_Porte, T. R. and P. M. Consolini (1991). "Working in practice but not in theory: theoritical challenges of 'high-reliability organizations'." *Journal of Public administrations*(1), Rochlin, G. I. and A. von_Meier (1994). "Nuclear power operations: a cross-cultural perspective." *Annual review of energy and the environment*(19): 153–187, Perin, C. (1998). "Operating as experimenting: synthesizing engineering and scientific values in nuclear power production." *Science, technology & human values* **23**(1): 98–128, Bourrier, M. (1999). *Le nucléaire à l'épreuve de l'organisation*. Paris, Presses universitaires de France, Journé, B. (1999). Les organisations complexes à risques: gérer la sûreté par les ressources. Etude de situations de conduite de centrales nucléaires. *Sciences de l'homme et de la société. Spécialité Gestion*. Paris, Ecole Polytechnique. **Thèse de doctorat**: 434, Rochlin, G. I. (1999). The social construction of safety. *Nuclear safety. A human factors perspective*. J. Misumi, B. Wilpert and R. Miller. London, Taylor & Francis: 5–23, Colmellere, C. (2008). Quand les concepteurs anticipent l'organisation pour maîtriser les risques: deux projets de modifications d'installations sur deux sites classés SEVESO 2. *Sociologie*. Paris, Université de technologie de Compiègne. **Thèse de doctorat**: 409.

by safety ethnographers seem hardly compatible with the constraints of the human factors assessment.

2.1. The Difficulties of Causal Analysis

Investigations following spectacular accidents often offer a wealth of lessons, especially when they are outstandingly well documented[203]. By reconstructing *a posteriori* the event chains, these investigations establish the causal relationships between safety and certain variables identified and thus increase knowledge.

However, several authors have raised doubts about the validity of these reconstructions. For example, Benoît Journé (1999) summarizes them thus: "The fact that the analyses always work backward from the end of the story to the assumed causes does not create objectivity. Knowing how the story ends influences the interpretations and even the attention directed to an element or other. When relating causes to the effects observed, investigators who are faced with a huge mass of different kinds of information are tempted to leave out elements that do not neatly fit the explanatory framework. Only those elements that explain the accident are highlighted. The issue is thus to know if the investigator will overinterpret and see causes that are actually not really causes."[204] Regarding this, Perrow (1994) wrote that an inspection conducted shortly before the Bhopal disaster had not identified any failures, whereas an inspection conducted *after the disaster* concluded that the accident was in fact unavoidable given the many failures it had found.

Nicolas Dodier's analyses[205], which are based on ethnographic investigations in industrial environments and which allowed Dodier to observe the reconstruction of occupational accidents by various experts, do not challenge the validity of reconstructions but they do underline their conventionality. Dodier distinguishes two accident analysis methods whereby the accident is viewed as a moral fact or as a failure. The first method focuses on solving a moral crisis created by the accident. To end the crisis, an accountable party must be identified. This search, often conducted within a legal framework, leads to a

---

[203] Shrivastava, P. (1987). *Bhopal: anatomy of a crisis.* Cambridge, Ballinger, Vaughan, D. (1996). *The Challenger launch decision: risky technology, culture, and deviance at Nasa.* Chicago, University of Chicago Press.

[204] p. 197.

[205] Dodier, N. (1994). "Causes et mises en cause. Innovation sociotechnique et jugement moral face aux accidents du travail." *Revue française de sociologie* **35**(2): 251–281, Dodier, N. (1995). *Les hommes et les machines. La conscience collective dans les sociétés technicisées.* Paris, Editions Métailié.

significant ontological distinction between objects and humans. On the other hand, when an accident is viewed as a failure, the accountable party, the cause, is not sought. Instead, an attempt is made to find a maximum number of elements involved in the event chains that led to the accident by placing humans and nonhumans on an equal footing[206]. It is this view of the socio-technical innovator, the network consolidator, that interests us, for it is how human factors experts look at incidents[207].

In describing the approach of socio-technical innovation, Dodier refers in particular to the properties of its associated tool – cause tree analysis – such as they are mentioned in a user's manual[208] or indicated by users. In particular, he says that "as a matter of principle, cause tree analysis never is the 'tree of all accident causes'. When an accident is treated as a failure, no tree is exhaustive."[209]

This idea is important for it is not without consequences on both *a posteriori* and *a priori* causal analysis. If it is impossible to exhaustively determine the consequences of an incident *a posteriori*, how then can all the event combinations that interweave humans and nonhumans and which may lead to an incident be found? This is a major risk management issue that restricts the

---

[206] Dodier states that this analysis type closely resembles the sociological theories of the actor-network, which are developed in particular at the Centre de sociologie de l'innovation (Center for the Sociology of Innovation) (cf. Akrich, M., M. Callon, et al. (2006). *Sociologie de la traduction: textes fondateurs.* Paris, Les presses de l'Ecole des mines). In the field of industrial safety, there are several proponents of the integration of humans and technical objects in risk analyses, one of whom is David Blockley (Blockley, D. I., Ed. (1992). *Engineering safety.* Berkshire, McGraw Hill, Blockley, D. I. (1996). Hazard engineering. *Accident and Design.* C. Hood and D. K. Jones. Abingdon, University College London Press: 31–39).

[207] This placement of the human and the nonhuman on an equal footing requires in-depth knowledge of the facility's socio-technical processes. For example, we saw that the Artémis incidents analysis required the human factors expert to work closely with the generalist engineer, who was trained in the various chemistry techniques used in the laboratory.

[208] A method published by the Institut pour l'amélioration des conditions de travail (Institute for the Improvement of Working Conditions), which is dependent on on the French trade union CFDT, Dodier, N. (1994). "Causes et mises en cause. Innovation sociotechnique et jugement moral face aux accidents du travail." *Revue française de sociologie* **35**(2): 251–281, p. 254.

[209] Ibid., p. 258.

effectiveness of safety strategies based on the forecasting of incident scenarios[210].

In a way, the difficulties of causal analysis that we have mentioned are not specific to risk. Paul Veyne discusses them in his essay on the epistemology of historical knowledge[211]. According to Veyne, facts are organized by the plot chosen by the historian, who shapes this organization and defines the causal relationships between facts.

"It is impossible to decide that one fact is historic and that another is an anecdote deserving to be forgotten, because every fact belongs to a series and has relative importance only within its series."[212] "Facts thus have a natural organization that the historian finds ready-made, once he has chosen his subject and it is unchangeable – The effort of historical work consists precisely in discovering that organization – causes of the 1914 War, the goals of the belligerents, the Sarajevo incident; the limits of the objectivity of historical explanations partly return to the fact that each historian succeeds in pushing the explanation more or less far. Within each subject chosen, this organization of facts confers on them a relative importance; in a military history of the 1914 War, a surprise attack on forward posts is of less importance than an offensive that filled the newspaper headlines; in the same military history, Verdun counts more than Spanish flu. Of course, in a demographic history the reverse will be true. The difficulties would begin only if one took it into one's head to ask whether Verdun or flu is absolutely more important from the point of view of History. Thus, facts do not exist in isolation, but have objective connections; the choice of a subject in history is free but, within the chosen subject, the facts and their connections are what they are and nothing can change that; historical truth is neither relative nor inaccessible, as something ineffable beyond all points of view, like a 'geometrical figure'."[213]

"Events compose a plot in which everything is explicable but unequally probable. ... events have causes but causes do not always have consequence – in

---

[210] Wildavsky, A. (1988). *Searching for safety*. New Brunswick, Transaction publishers, Hood, C. and D. K. C. Jones, Eds. (1996). *Accident and design. Contemporary debates in risk management*. Abingdon, University College London Press, Journé, B. (1999). Les organisations complexes à risques: gérer la sûreté par les ressources. Etude de situations de conduite de centrales nucléaires. *Sciences de l'homme et de la société. Spécialité Gestion*. Paris, Ecole Polytechnique. **Thèse de doctorat**: 434.

[211] Veyne, P. (1971 (1996)). *Comment on écrit l'histoire?* Paris, Seuil.

[212] Ibid., p. 37.

[213] Ibid., p. 51.

short, the chances of different events happening are unequal."[214] "The choice of plot decides supremely what will or will not be causally relevant; science can make all the progress it wants, but history clings to its fundamental option, according to which the cause exists only through the plot. For such is the real meaning of the notion of causality. Let us suppose that we have to state the cause of a car accident. A car has skidded after braking on a wet, cambered road: to the police, the cause is excessive speed or worn tires; to the Department of Bridges and Highways, the high camber; to the head of a driving school, the law, not understood by learners, which requires that the distance for braking increases more than proportionally with the speed; for the family, it is fate, which willed it should rain that day or that this road existed for the driver to come and be killed there."[215]

Some approaches predicated on an ideal model make it possible to overcome the difficulties posed by the causal analysis: several "lines of defense", "organizational barriers" are meant to make it possible to avert incidents (or mitigate their consequences) without needing to explain them *a priori*. Furthermore, as we have seen in the case of the Artémis incidents, the causal analysis does not obviate the use of a reference model, which consists of a guide for reconstructing the event chains. It shapes "the plot" and makes it possible to arrive at conclusions and make recommendations. In addition to appearing indispensable, the use of a reference model is backed by academic research, notably that of James Reason.

2.2. The Organizational Barriers to Safety Seen Through Reason's Model

In the 1960s James Reason began working in the field of aviation and particularly cockpit ergonomics. His career led him to analyze and theorize human error[216]. He summarized his research in his 1990 book titled "Human Error" at the end of which he makes a case for taking better account of organizational errors in the analysis of industrial accidents. His discussion is based on basic notions and a list of factors integrated in a model that he fine-tuned in the 1990s and which, despite the criticism it has received, has become the dominant paradigm for researchers and influencers in the field of risk management.

---

[214] Ibid., p. 198.

[215] Ibid., pp. 226–227.

[216] Reason (1993) suggests dividing errors into three classes: skill-based errors, rule-based errors and knowledge-based errors. Added to these are violations, which are a particular type of error in that they are voluntary.

Chernobyl is one of the many accidents that Reason studied. "The ingredients for the Chernobyl disaster were present at many levels. There was a society committed to the generation of energy through large-scale nuclear power plants. There was a system that was hazardous, complex, tightly coupled, opaque and operating outside normal conditions. There was a management structure that was monolithic, remote and slow to respond. There were operators who possessed only a limited understanding of the system they were controlling, and who, in any case, were set a task that made violations inevitable."[217]

The following comments illustrate the broader message that he subsequently defended: "All man-made systems have within the seeds of their own destruction, like 'resident pathogens' in the human body. At any one time, there will be a certain number of component failures, human errors and 'unavoidable violations'. No one of the agents is generally sufficient to cause a significant breakdown. Disasters occur through the unseen and usually unforeseeable concatenation of a large number of these pathogens."[218] Reason's observations follow the analyses of Barry Turner (1978), who emphasized the multi-causal and non-linear nature of accidents and, Charles Perrow (1984), who characterized certain high-risk organizations by their complexity and tight coupling[219].

Chernobyl, TMI, Bhopal, Challenger and many other disasters led Reason to put his work on human error into perspective, as illustrated by this excerpt from "Human Error": "Rather than being the main instigators of an accident, operators tend to be the inheritors of system defects created by poor design, incorrect installation, faulty maintenance and bad management decisions. ... There is a growing awareness within the human reliability community that attempts to discover and neutralize these latent failures will have a greater beneficial effect upon system safety than will localized efforts to minimize active errors. To date, much of the work of human factors specialists has been directed at improving the immediate human-system interface (i.e., the control room or *cockpit*). While this is undeniably an important enterprise, it only addresses a relatively small part of the total safety problem, being aimed primarily at reducing the 'active failure' tip of the causal iceberg. One thing that has been profitably learned over the past few years is that, in regard to safety

---

[217] Reason, J. (1987). "The Chernobyl errors." *Bulletin of the British psychological society*(40): 201–206, p. 206.

[218] Ibid., p. 206.

[219] These notions are discussed in the following subsection.

issues, the term 'human factors' embraces a far wider range of individuals and activities than those associated with front-line operators such as pilots"[220]. The first version of Reason's model appeared in this book.

Reason's distinction between active errors and latent conditions forms the basis of his model, whose second "linchpin" is a concept that we have already mentioned – defense in depth, considered in a broad sense as a layering of "barriers". Figure 1 shows one of the representations Reason proposed for his model[221].

The model of accident causation consists of three basic elements: hazards, defenses and losses. It "seeks to link the various contributing elements into a coherent sequence that runs bottom-up in causation and top-down in investigation."[222]

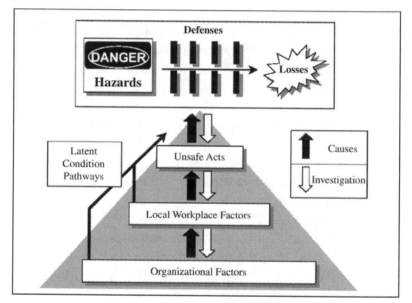

Figure 1: A representation of Reason's model of accident causation (1997).

Reason then explains in detail the contributing elements of accident causation, i.e. those that prevent the lines of defense from containing the accident. He distinguishes organizational factors from local factors:

---

[220] Reason, J. (1993). *L'erreur humaine.* Paris, Presses universitaires de France, p. 240.

[221] Reason, J. (1997). *Managing the risks of organizational accidents.* Aldershot, Ashgate Publishing Limited, p. 17.

[222] Ibid., p. 16.

"The causal story starts with the *organizational factors*[223]:

- *strategic decisions,*
- *generic organizational processes – forecasting, budgeting, allocating resources, scheduling, planning, communicating, managing, auditing and the like. (...)*

The consequences of these activities are then communicated throughout the organization to individual workplaces – control rooms, flight decks, air traffic control centers, maintenance facilities and so on – where they reveal themselves as factors likely to promote unsafe acts. [These factors] include

- *undue time pressure,*
- *inadequate tools and equipment,*
- *poor human-machine interfaces,*
- *insufficient training,*
- *undermanning,*
- *poor supervisor-worker ratios,*
- *low pay,*
- *low status,*
- *macho culture,*
- *unworkable or ambiguous procedures,*
- *poor communications and the like.*

Within the workplace, the *local factors* combine with natural human tendencies to produce errors and violations – collectively termed 'unsafe acts' – committed by individuals and teams at the 'sharp end' or the direct human-system interface. Large numbers of these unsafe acts will be made, but only very few of them will create holes in the defense."[224]

The representation shown in Figure 2 is behind the model's success. The successive barriers have breaches, some due to active errors (unsafe acts) and others to latent conditions (organizational and local risk factors).

---

[223] Reason's model is quite similar to that of the human factors experts that we built from the five assessments. However, Reason gives additional factors, which are underlined here.

[224] Reason, J. (1997). *Managing the risks of organizational accidents.* Aldershot, Ashgate Publishing Limited, pp. 16–17.

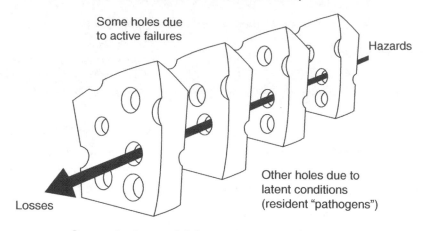

Successive layers of defences, barriers and safeguards

Figure 2: The "Swiss Cheese model".

Reason's model has inspired many risk management experts in many sectors. The risk factors identified by Reason trivially give the safety factors that could be included in action plans to prevent potential breaches in defense in depth.

Although Reason refers to cultural factors[225], the conception of safety and its control conveyed by his model is based primarily on a layering of component and organizational systems meant to supervise, guide and help operators, and avert or contain unsafe acts. It contains many of the previously listed variables used by the human factors experts – notably quality of information, of HMIs, of procedures, of working conditions and of communication methods. The underlying logic is to set up systems without needing to explain in principle the event chains and thus without the causal analysis.

One can thus imagine one of the model's criticisms – the weakness of the cognitive links between these systems and safety – which Reason himself expressed several times:

- *"The deterministic causal connection between latent conditions and accidents cannot easily be identified (particularly before the event),"*[226]

- *"A condition is not a cause, but it is necessary for a causal factor to have an impact."*[227]

---

[225] Reason (1997) points out that organizational factors "are drawn and carved by corporate culture, implicit attitudes and informal rules." (p. 16).

[226] Reason, J., E. Hollnagel, et al. (2006). Revisiting the "swiss cheese" model of accidents. Brétigny-sur-Orge, Eurocontrol, p. 13.

[227] Ibid., p. 7.

• *"The [Swiss Cheese model] does not provide a detailed accident model or a detailed theory of how the multitude of functions and entities in a complex socio-technical system interact and depend on each other. That does not detract from its value as a means of communication, but may limit its use in analysis and as a support for proactive measurements."*[228]

2.3. The Ethnographic Approaches to Safety

Other researchers have taken a different route in studying safety and, more generally, the reliability of high-risk organizations. Their initial questions are no longer "What are the causes of such-and-such accident or such-and-such series of accidents?" or "What system can be put in place to prevent these accidents from recurring?", but rather "What features allow some organizations to achieve a very high level of safety, exemplified in particular by the small number of accidents?"

In the 1980s researchers in the Berkeley high-reliability organizations (HRO) project were among the first to frame and examine these questions. They have since been supplanted by others, most notably the French academics Mathilde Bourrier and Benoît Journé and the American cultural anthropologist Constance Perin. Although their findings depended in particular on their academic discipline, the common feature of these works of is that it they use a material generally obtained from in-depth field investigations. The HRO group's initial questions and observations are a critique of the findings of Charles Perrow, who had analyzed TMI and other accidents, and classical organizational theories. The ethnographic approaches to safety led their authors to formulate concepts that, although novel, were harder to manipulate; this is one of their limits.

The HRO group's initial premises challenge those of Perrow's normal accident theory (1984), to which Scott Sagan contributed (1993). Perrow sees high-risk organizations as systems that have specific properties and bring about their unreliability. The primary property is the presence of complex interactions, which Perrow defines as those of unfamiliar sequences, or unplanned and unexpected sequences, and either not visible or not immediately comprehensible by operators. The particular nature of the coupling between

---

[228] Ibid., p. 21.

technology and the organization of work, which he calls "tightly coupled"[229] is also a factor of unreliability. This property has specific implications – the sequences in the process are invariant, there is only one way of getting something done and there are few substitution possibilities for obtaining supplies, equipment and humanpower. Put together, these constraints limit an organization's substitution capabilities. System redundancies are designed in advance, limiting an organization's ability to cope with unexpected events. Perrow also notes that although, on the whole, social organizations are rather loosely interconnected systems, high-risk systems are tightly interconnected due to the technology they use. Design flaws are thus irreversible since the technology used requires an unattainable level of perfection in the sharing of information. In short, accidents are normal, in the sense that they are unavoidable, and reliability is beyond reach.

The members of the HRO group challenge Perrow's conclusions, asking how is it that, despite the disaster potential revealed by normal accident theory, the statistics do not show more accidents? "Like Perrow, they do not believe individuals are sufficiently rational to individually master the complexity of high-risk technologies. However, unlike Perrow, they think that the organization has the means to make up for [the] individual limitations and achieves levels of reliability never before attained by any other industrial system. What Perrow calls 'high-risk technologies' must, at the same time, be called 'high-reliability organization'."[230] This controversy, which is also fueled by the fact that Perrow considers certain reliability criteria adopted by the HRO group's researchers to be a source of unreliability (particularly redundancy), continues to strongly shape debate and the views of many researchers.

Furthermore, the group's initial observations of four different areas of activity (air traffic control, electric utility grid management, nuclear power plant operations and the operation of an aircraft carrier) also contradict the classical findings of organizational theories. In a retrospective analysis, Gene Rochlin, one of the HRO project's initial participants, states that "our original assumptions and trial hypotheses had, if anything, been too timid and our scope too narrow."[231] In particular, Rochlin refers to a difference with learning theories: "The classical organizational theories develop a broad set of

---

[229] As opposed to organizations such as the universities considered by Karl Weick to be loosely coupled (cf. Weick, K. (1976). "Educational organizations as loosely coupled systems." *Administrative science quaterly* **21**(1): 1–19).

[230] Journé, 1999, p. 63.

[231] Rochlin, G. I. (2001). Les organisations à "haute fiabilité": bilan et perspectives de recherche. *Organiser la fiabilité*. M. Bourrier. Paris, L'Harmattan: 39–70, p. 41.

institutional learning models, from formal rationality, for which learning occurs through trial and error, to the more informal garbage can model. Most of these approaches are based on an organization that is loosely interdependent on its environment, which makes the consequences of the learning process acceptable. Such was not the case [of the] organizations [studied], which shared the conviction that their first major error would probably be the last."[232]

The findings reported[233] by the HRO group's researchers and by those who have examined safety "as it is being developed" were also markedly different from those established by the proponents of a barrier-based approach.

In several of their publications, the representatives of the HRO group place particular emphasis on *the importance of informal aspects and work atmosphere*, which they consider to be factors that contribute to maintaining facilities in a safe operational state. Here are a few examples:

- *"An analysis of extensive video and audio recordings of conversations between controllers and pilots reveals many rules and interactions that are not formalized, but are meant to nurture communication and cooperation."*[234]

- *"The unwritten social rules of the plant must allow an individual to discover and draw attention to an error or problem without being blamed for it, and must guarantee that any individual's concerns are taken seriously and will be responded to."*[235]

- *"Following rules and procedures rigorously is not the only guarantee of safety; curiosity and professional initiative are also required. In all, safety depends on multiple kinds of knowledge, rationalities and values."*[236]

Rochlin and von Meier use the concept of *culture* by departing from the definitions used in the requirement guidelines or other analyses of the literature: "We take culture to be neither a holistic narrative or philosophical text, nor a formal construct or property subject to quantifiable analysis as it is in studies of 'organizational' culture, nor even a means or mechanism for capturing a wide variety of performative variables as it is in most studies of 'safety culture'. Rather, it is a heuristic device for exploring and describing historically and

---

[232] Ibid., p. 41.

[233] They are underlined on the following pages.

[234] Rochlin, G. I. (1999). The social construction of safety. *Nuclear safety. A human factors perspective.* J. Misumi, B. Wilpert and R. Miller. London, Taylor & Francis: 5–23, p. 15.

[235] Rochlin, G. I. and A. von_Meier (1994). "Nuclear power operations: a cross-cultural perspective." *Annual review of energy and the environment* (19): 153–187, pp. 167–168.

[236] Perin, C. (1998). "Operating as experimenting: synthesizing engineering and scientific values in nuclear power production." *Science, technology & human values* **23**(1): 98–128, p. 100.

socially determined behavior. To study culture in this light is to search for the often masked constitutive rules of collective behavior, to tease out the underlying framework that shapes the way in which internal as well as external social relations and social interactions of organizations are constructed."[237] In general, they condemn prevailing, traditional and "rationalistic" approaches, which they consider inadequate — if not dangerous — for characterizing safety and understanding its determining factors:

- *"Many analysts of, and consultants to, the nuclear industry, including, in many cases, on-site as well as off-site regulatory personnel, are grounded in approaches that are fundamentally rationalistic .... Ritualization ..., expressions of confidence that are not grounded in empirically determinable observables are all too often disregarded ...."* [238]

- *"My claim is that nuclear power plant operations are experimental systems that intentionally and unintentionally produce invaluable knowledge, including surprises, about their safety status. If this information is unnoticed, ignored, or misinterpreted, the consequences can be devastating."* [239] *"What these people know, abductively or otherwise, is dismissed by an epistemological caste system that does not value knowledge that is specific, home grown, and based on experience."* [240]

The members of the Berkeley HRO project are also known for proposing *a list of characteristics of high-reliability organizations*: "Although we found many differences that were of considerable use to us in arraying the organizations along different axes ... what was most striking were the many similarities and commonalities .... These we have summarized in many ways that go far beyond the usual notion of a 'culture of reliability', including, inter alia:

- *flexible delegation of authority;*

- *structure under stress (particularly in crises and emergency situations);*

- *respect for, and nurture of, the skill and dedication of operators and workers at all levels;*

- *a system of rewards for reporting and discovering error not just even, but*

---

[237] Rochlin, G. I. and A. von_Meier (1994). "Nuclear power operations: a cross-cultural perspective." *Annual review of energy and the environment*(19): 153–187, p. 159.

[238] Rochlin, G. I. (1999). The social construction of safety. *Nuclear safety. A human factors perspective*. J. Misumi, B. Wilpert and R. Miller. London, Taylor & Francis: 5–23, p. 18.

[239] Perin, C. (1998). "Operating as experimenting: synthesizing engineering and scientific values in nuclear power production." *Science, technology & human values* 23(1): 98–128, p. 106.

[240] Ibid., p. 115.

especially, one's own;

- *a mix of welcome for, and resistance to, technical and organizational change that is based almost entirely on thoughtful evaluation of their short- and long-term effects on organizational reliability and performance.*[241]

In a presentation summarizing the work of the Berkeley HRO group, Sagan (1993) identified other criteria, including:

- *redundancy of channels of decision-making,*

- *cross-checks,*

- *continuous education and training,*

- *agreement of the organization's members on the organization's end goals.*

By choosing to observe the behaviors of control room operators during normal operating conditions, Journé adopted a methodology similar to that of the Berkeley HRO group. Journé's concept of *organizational arrangement*[242] and Karl Weick's concept of *sensemaking*[243], which Journé uses, firmly establish Journé's approach in the field of the cognitive approach to organizations. For Journé, safety depends not just on the ability to control foreseeable incident and accident conditions but also the ability of the control room to manage, on a daily basis, normal yet unexpected conditions. His thesis argues for "pragmatic complexity management, which consists in ensuring that organizational arrangement is not left defenseless against unexpected events."[244] Like the

---

[241] Rochlin, G. I. (2001). Les organisations à "haute fiabilité": bilan et perspectives de recherche. *Organiser la fiabilité*. M. Bourrier. Paris, L'Harmattan: 39–70, pp. 46–47.

[242] "Organizational arrangement is an agent (an agency) that is a mix of various resources, the most notable of which are human resources, material resources and symbolic resources." (Girin 1995)

[243] "Process by which each individual attempts to construct his own area of sense, his "reality", by extracting meaningful patterns from experiences and life situations." (Vidaillet, B., Ed. (2003). *Le sens de l'action. Karl E. Weick: sociopsychologie de l'organisation*. Paris, Vuibert, p. 177). "Despite its proximity to other explanatory mechanisms, such as comprehension, interpretation and attribution, sensemaking is distinguished from them by the fact that it does not assume the existence of a pre-existing sense that would be waiting to be discovered. Rather, sensemaking suggests the construction – the *enactment* – of a sense that incorporates the subject and its environment in a single interactive process." (Journé 1999, p. 153)

[244] Ibid., p. 415.

members of the HRO project and other researchers[245], Juré makes a case for implementing a resilient approach to safety (develop abilities to cope with and bounce back from unexpected events).

Without challenging the relevance of the findings and methods of this research, the determining factors of safety highlighted, and particularly the informal cultural aspects, do not appear easy to consider in a prescriptive framework, for the findings are not yet robust enough: "The challenge for the group's work – indeed, for all safety analysis – is to take such examples, together with the work of others who have come to similar conclusions, and to try to construct from them a definition of operational safety that is analytical, robust, predictive, and not just contingent on past performance (or luck); and to be able to decide what is and what is not valuable and important for the social construction of safety, without marginalizing those who speak for experience, history, culture, social interactions, or other difficult-to quantify sociocultural and socioanthropological variables."[246] The same holds true for the concept of resilience; an international team of researchers is currently tackling the challenge to make it "workable"[247]. Although these researchers admit that this may be a promising avenue, the operational benefits are not yet forthcoming.

Nevertheless, the criteria identified by the Berkeley HRO group are "positive" findings, the "bricks" of a high-reliability organization. However, they refuse to advocate a particular organizational model based on this criteria. According to them, there is no such thing as a high-reliability organization model. This is also the opinion of Mathilde Bourrier (1999). Her investigations at four nuclear sites in France and the USA during outage show that the aforementioned criteria that would distinguish high-reliability organizations from others "are merely partially identifiable in the four systems ... studied. ... Take the example of the criterion on the agreement of goals. In Nogent's case, the criterion does not apply. Nogent's management and operators disagree about the organization's future. Logically, this should exclude Nogent from the category of [high-reliability organizations]. And yet, despite this disagreement about the goals – which leads to a lack of social cohesion within the plant, which itself is reflected by a loss of

---

[245] Wildavsky, A. (1988). *Searching for safety*. New Brunswick, Transaction publishers, Collingridge, D. (1996). Resilience, flexibility, and diversity in managing the risks of technologies. *Accident and design: contemporary debates in risk management*. C. Hood and D. K. C. Jones. London, UCL Press limited, Hollnagel, E. and E. Rigaud, Eds. (2006). *Proceedings of the second resilience engineering symposium. 8–10 November 2006, Antibes*. Paris, Les Presses de l'Ecole des Mines, Hollnagel, E., D. D. Woods, et al. (2006). *Resilience engineering*. Hampshire, Ashgate.

[246] Rochlin, 1999, pp. 17–18.

[247] Known as the Resilience Engineering Network , Erik Hollnagel is one of its members.

points of reference and a phenomenon of anomie – it cannot be concluded that Nogent is 'unreliable'"[248].

2.4. Conclusion: The Lessons of the Human Factors Assessment

The conception of safety advocated by safety ethnographers is different from the conception predicated on defense in depth and Reason's model, whose principles are quite similar to those used by the human factors experts[249].

The findings of this research on safety "in action" hardly appear prescriptible. Indeed, we cite as an example "an intergroup tension between units organized around different professional skills, approaches or training" underscored by Rochlin and von Meier: "Such tensions are difficult to read, let alone interpret, without complete familiarity with the social and cultural dimensions of the plant and its operating environment. In some cases intergroup rivalries may be a warning sign of difficulties or tensions within the plant. In others (including several my colleagues and I have observed) they are part of the dynamic process of making sure that the right group owns an impending error or difficult task. In still others, the conflict may be formally or culturally ritualized, and therefore not just intersubjective but also structural. The temptation to view such tensions as interfering with operational safety, or as being largely exogenous to it, is great, but the pattern we have observed in plants that are operating well and safely is one of a balance between collaboration and conflict, and between mutual respect and occupational chauvinism, a balance that serves the dual purpose of maintaining group self-identity and cohesion and providing a necessary channel for redundancy and backup."[250] Although the notion of balance between collaboration and conflict undoubtedly is relevant, it hardly seems feasible and consequently not easily integratable in safety assessments[251].

Although we can only encourage the experts to continue to follow the findings in the "alternative" scientific literature, an analysis through comparison with the reference model[252] or a causal analysis – despite the limitations of

---

[248]Bourrier, M. (1999). *Le nucléaire à l'épreuve de l'organisation*. Paris, Presses universitaires de France, pp. 257–258.

[249] However, the human factors experts seek to understand "real life" through case studies.

[250] Rochlin, 1999, p. 17.

[251] Remember the reaction of the head of the SEFH when one of the members of the advisory committee spoke of the importance of distributed skills: "We've conducted many interviews and made many observations but I don't think that an IRSN review can see all that. That's one of its limitations."

[252] The human factors experts might possibly complete the reference model by drawing from the factors cited by James Reason.

forecasting – seem better suited to the prescriptive task of the assessment. Lastly, note that despite an apparently weak causal link with safety, the model's human and organizational factors are identified within a community of researchers as being safety factors.

3. Institutional and Cognitive Rationality of the Human Factors Experts

Given the weakness of the established knowledge in the field of human factors, how can the choice of one analysis type over another be justified? Why favor one argument over another? Can rationalities likely to explain the experts' choices be identified?

At the end of his rationality that made reference in particular to the discussions of Max Weber and Herbert Simon, the sociologist Raymond Boudon concludes that "no formal definition can explain [the notion of rationality] and it can only be given a deictic definition: a behavior or belief Y of which it can be said that 'subject X has good *reasons* to do (believe) Y, because...' is *rational*; a behavior or belief Y of which it cannot be said that 'subject X has no reason to do (believe) Y, *but...*' is *irrational*"[253]. We would like to make a list of the "good reasons" that help to explain the experts' choices. This approach, which is characteristic of a tradition in management research[254], will lead us to explain the experts' objectives and the constraints they face. These objectives and constraints make up what we will call the "central problem of the assessment". We will show that an assessment based on cognitively weak arguments can enable the expert to achieve his objectives and, at the same time, offer him the opportunity to overcome the difficulties of conducting a causal analysis. We will reveal the existence of a rationality that justifies the use of arguments whose cognitive limitations we have pointed out. The nature of the "good reasons" that characterize this rationality led us to describe it as "institutional". Additional requirements may, however, call for conducting a causal analysis, which is more in keeping with the proponents of a cognitive rationality. In a manner of speaking, this amounts to favoring conviction over compliance.

---

[253] Boudon, R. (1990). *L'art de se persuader des idées douteuses, fragiles ou fausses.* Paris, Librairie Arthème Fayard, p. 374.

[254] Berry, M., J.-C. Moisdon, et al. (1978). "Qu'est-ce que la recherche en gestion?" *Informatique et gestion* **108–109**, Berry, M. (1983). Une technologie invisible? L'impact des instruments de gestion sur l'évolution des systèmes humains. Paris, Centre de recherche en gestion, Ecole Polytechnique: 98, Moisdon, J.-C. (1984). "Recherche en gestion et intervention." *Revue française de gestion.* 61–73, Riveline, C. (1991). "Un point de vue d'ingénieur sur la gestion des organisations." *Gérer et comprendre.* 50–62.

3.1. The Central Problem of the Assessment

In order to explain an expert's choices, we suggest viewing his assessment as a problem-solving strategy – consisting primarily of achieving objectives – that must meet constraints. This strategy will be characterized by the analysis type used the most.

The change that led to incorporating human factors in the institute's assessment activity marked a key moment from which the human factors experts began conducting their work as part of an assessment and thus became obliged to make recommendations. Furthermore, these recommendations had to align the various viewpoints of the assessment's stakeholders. The chances of the expert's proposed assessment achieving satisfactory rhetorical effectiveness had to be good. It must also be said that the expert's assessment had to stay within certain institutional constraints. The pre-eminence of such objectives, however minor, may come as a surprise. Yet, it is the result of our immersion, observations and interviews.

First objective: recommend

As we have already said several times, like any expert, the human factors experts were expected to make conclusions. Moreover, this is the first of the four principles of the head of the SEFH that we explained in chapter 3.

Associated with a principle of continual safety improvement demanded by the representatives of the external review institutions (IRSN, ASN, members of the advisory committee), this objective of conclusions became an objective of recommendations. "I don't see how I couldn't make recommendations in my assessment. That would mean that everything is perfect. However, we're dealing with continual improvement here." (a human factors expert, interview of November 21, 2005). This was the expert's first objective.

Second objective: achieve satisfactory rhetorical effectiveness

Making recommendations is a precondition for solving the central problem of the assessment, but it is not sufficient. These recommendations must be seen as sufficiently legitimate. This "legitimation process" depends on the assessment type. In the case of an assessment presented before the advisory committee, the following conditions must be met:

- *The expert makes recommendations;*
- *Most of these recommendations are viewed as legitimate within the IRSN;*
- *A portion of these recommendations (part A) consists of recommendations accepted by the licensee (changed to commitments or stands/actions). The "number of recommendations accepted by the licensee/number of*

recommendations from part A" ratio is rather high;

- *The remaining portion (part B) consists of recommendations discussed during the advisory committee meeting. The "number of recommendations chosen by the advisory committee/number of recommendations from part B" ratio is rather high.*

When an assessment is not presented before the advisory committee, the licensee is not involved in the legitimation process. The adequate preconditions for achieving the objective are as follows:

- *The expert makes recommendations;*

- *The "number of recommendations viewed as legitimate within the IRSN and by the ASN's regional representatives/number of recommendations proposed by the expert" ratio is rather high.*

When an assessment meets the central problem's two objectives, we consider it to be rhetorically effective. It can be noted that these objectives are institutional in nature. It is the institutional framework of nuclear safety that motivates the experts to make recommendations and the quality of their recommendations is judged by representatives of the institutions within this framework.

Fulfilling these two objectives is no simple matter. Once an assessment has been completed and recommendations made, there remains the task of persuading the IRSN, the licensee and the advisory committee while taking the following information into account:

- *Most of the IRSN's experts are technical specialists who are not always familiar with the benefits of the reference model that forms the basis of many human factors experts' recommendations;*

- *If implemented, the aforementioned recommendations of the human factors experts result in one of two things: the establishment or consolidation of a human or organizational factor or the performance of studies. In short, a recommendation nearly[255] always means that the licensee will be obligated to make investments and, as a result, may want to oppose the expert's recommendation;*

- *Because the licensee is also represented by the members of the advisory committee, it has the opportunity to oppose the experts' conclusions as well.*

Thus, even when the expert is able to evaluate the benefits resulting from his recommendations (reduced losses from potential incidents), succeeding in a rhetorically effective assessment is no easy task.

---

[255] In other cases, the expert may request that a study in progress be passed on.

Institutional constraints

While the expert must recommend and align, he must also propose an assessment that satisfies the following constraints:

- *The expert's assessment fulfills a request. Of course, our accounts show that the human factors experts could take part in formulating such a request. Nevertheless, this request is set in an institutional context not fully controlled by these experts (e.g. the ASN's demand for regular cross-disciplinary assessments of the nuclear power plant fleet, the demand of the ASN and the generalist engineers for a human factors contribution to safety reviews of CEA facilities)*

- *The assessment process is bounded by time constraints. As illustrated by our cases, each stage of the assessment is limited in terms of time. Although deadlines can be renegotiated, the importance of meeting them is regularly brought up. Furthermore, especially regarding data collection, the expert must yield to the constraints of the assessment's other stakeholders. A good example of this is when the occupation studies for the skills management assessment could not be started until the EDF sites had been selected.*

- *The assessment must meet formal constraints (e.g. number of pages, compliance with institutional rhetorical forms). It is not uncommon for the members of the advisory committee to give their opinion on such matters as the report's volume or the quality of the rapporteur's presentation.*

Like the aforementioned objectives, these constraints can be described as "institutional" in that they are exerted by the various assessment stakeholders (experts, licensees, the ASN, members of the advisory committee).

Strengths and effects of the objectives

Before continuing our modeling effort, we would like to add a few comments on the objectives explained. This is in no way to say that the experts do not aim to improve safety. Nevertheless, what our accounts make clear is that improving safety involves tests of judgment. It is these stages that most shape the experts' behavior, that must ensure that they are successfully completed and, accordingly, propose an assessment that will fulfill the two aforementioned objectives. Admittedly, rhetorical effectiveness is a minimum objective[256] for the IRSN's experts. Nevertheless, it is essential and independent of the expert's hierarchical level and specialized field. By way of example, here is a comment of a generalist engineering department head after the Minotaure review

---

[256] We will see later on that the experts can take other objectives into consideration.

assessment is an example: "The review wasn't easy, but I'm pleased with the findings. We turned many recommendations into commitments and eighteen of the assessment's twenty draft recommendations were changed to recommendations. We had the advisory committee's support." (interview of March 29, 2006)

In keeping with classical findings in management research, we point out that these objectives, which can be incorporated in simple indicators, are not without collateral effects. "It's a problem at the IRSN. Each expert accentuates his own particular field. Because the experts make recommendations right and left, there are lots of them. The problem is that the recommendations are not prioritized." (a member of the advisory committee speaking about the Minotaure review assessment, interview of June 2, 2006). "It's disastrous that the experts think that it's not good if they don't make recommendations." (a member of the advisory committee, interview of June 9, 2006)

The identification of "local rationalities", out of step with an assumed traditional rationality, is another task that the aforementioned management researchers tackled. Thus, like our forerunners, by revealing the "good reasons" used by the experts to justify their choices, we will reveal the attributes of an institutional rationality out of step with cognitive rationality.

3.2. Cognitively Weak Arguments as a Solution to the Central Problem of the Assessment

We are going to show that using cognitively weak arguments can make it possible to solve the central problem of the assessment by allowing the expert to satisfy constraints and at the same time attain the aforementioned objectives.

Satisfaction of constraints

Satisfying formal constraints depends on neither the type of analysis nor the type of argument used. Analysis through comparison with the reference model (or with a portion thereof) is therefore compatible with these constraints. These processes – which, as we have seen, result in weakly cognitive arguments – also make it possible to satisfy the two other constraints of the central problem: fulfilling the request and satisfying the time constraints. The expert can use the analysis through comparison with the reference model for assessments of all types with the exception of incident analyses for which the causal analyses are "naturally" used. The expert did so during the Minotaure review and skills management assessments. The reference model, which focuses on the organization as a whole, is even relatively suited to safety reviews and cross-disciplinary assessments and their initial questions that can be vague or general (e.g. "give an opinion on the consideration given to human factors in the safety review approach").

The analysis through comparison with the reference model requires less effort than a causal analysis. Indeed, as we saw in the both the cases of the Artémis and the Minotaure (rod fabrication) assessments, the causal analysis requires making an effort to understand the technical processes and the hazardous activities. This effort must be made by the human factors expert, who must work at the fringes of his area of expertise and must therefore consider working closely with generalist and specialist engineers, provided they are willing to do so (they were not in the case of the skills management assessment). In principle, such collaboration must be formalized by the institute's procedures and may increase the number of proofreading runs.

Thus, in most cases, the analysis through comparison with the reference model allows the expert to satisfy the constraints imposed on him. But do these choices allow him to propose recommendations and win over his audience?

Possible fulfillment of the objectives

As our three accounts show, recommendations are a virtually immediate outcome of the analysis through comparison with the reference model. Recommendations are in fact integrated in the analysis because the comparison of the data and the model shows a gap that must be bridged. "No formalization of the discussion on skills management", "No formalization of operating experience feedback" (Minotaure review assessment) and the poor quality of certain skills management systems, particularly for the mechanical operations supervisor (skills management assessment), are three examples.

For all that, can these recommendations lead the assessment stakeholders to a consensus? Sometimes. The three examples that we just mentioned led to commitments on the part of the licensee. The recommendations were validated internally and then submitted to the licensee, which accepted them. Also, the Minotaure review and skills management assessments cannot be criticized for their lack of rhetorical effectiveness. In the first case, nearly all the recommendations were accepted by the generalist engineers then changed into commitments by the licensee. In the second case, give or take a few details, the few recommendations that were not changed to stands/actions were made into recommendations of the advisory committee. Thus, an assessment with recommendations that are justified by cognitively weak arguments can be a solution to the central problem of the assessment. We have explained an institutional rationality, in that we revealed the "good reasons", of an institutional nature, that incentivize the expert to use this type of justification. Furthermore, this institutional rationality is in part legitimate within safety institutions. "The licensee shall establish the demonstration of safety" is one of the operating principles of technical dialogue institutions. The interpretation of this principle by some experts results in a mere evaluation of the data provided, the poor quality of which is sufficient to justify a recommendation. Furthermore, as we have seen, some experts are convinced of the relevance of the reference model. In addition to being used by many researchers, the importance of its component human and organizational variables was shown by the analysis of many industrial accidents, including TMI and Chernobyl.

However, although the objectives making up the central problem are shared by all, there may be others. Our accounts illustrate that some experts did not content themselves with just the attributes of this institutional rationality. In the next section we will try to explain why.

3.3. Using the Causal Analysis to Fulfill Additional Requirements

Although such is not always the case, the expert has "good reasons" to think that, *a priori*, the licensee will assume a challenging attitude. As a result, the proofreading runs result in further justifying the recommendations in the hopes of achieving a satisfactory level of rhetorical effectiveness. What is more, cognitive effectiveness may be a concern for experts anxious to improve the knowledge in the assessment described as weak. Both of these additional requirements can be met by the causal analysis.

Anticipate likely opposition by the licensee

As we have said, a recommendation nearly always means that the licensee will be required to make significant investments. One member of the advisory committee told us about another type of cost. "As the licensee sees it, the assessment is recommendation machine that leads to requests from the ASN. It interprets each recommendation as a constraint because it will have to account for things." (interview of March 30, 2006)

Without throwing doubt on whether they indeed do uphold the principles they spotlight in their corporate communications ("safety is our top priority"), nuclear utilities obviously have financial interests. Thus, although an incident may cost much more than improving a procedure, implementing a training program or conducting a study[257], financial constraints can place the licensee and the expert at loggerheads. The accounts of the meetings held during the skills management assessment show just how much opposition the experts can face when presenting their conclusions.

---

[257] It is generally estimated that an unavailable medium-sized nuclear unit leads to "losses" of €1 million a day.

The desire to know: for a cognitively effective assessment

In addition to the objective of rhetorical effectiveness, there may also be an objective of cognitive effectiveness. Several experts in the SEFH, including its head, want to enhance knowledge.

Some have PhDs or are former researchers, while others participate in scientific colloquia, write research articles or organize seminars at which policies and approaches are discussed. In short, several experts feel that it is their vested duty to enhance knowledge. Moreover, research figures prominently in the institute's presentations. "The IRSN is the French public service expert in matters involving nuclear and radiation risks."[258] A mark of the department's place in the academic community is the fact that it hosts PhD students and is represented in several international research groups on human factors in nuclear safety.

Furthermore, as part of their assessment activity, some experts want to test the reference model's organizational factors, challenge certain presuppositions and establish robust ties with safety. The experts may feel uncomfortable about using statements that they themselves are not sufficiently convinced of to persuade the licensee. The SEFH by and large thus has an objective of cognitive effectiveness.

Rhetorical and cognitive effectiveness of the causal analysis

To attempt to achieve a high level of both rhetorical and cognitive effectiveness requires demonstrating the causal link between the recommendation and safety. The strategy to be used to meet these new requirements is predicated on the causal analysis.

Indeed, merely shifting away from the reference model may not make it possible to win over an inflexible licensee. The licensee may bring up the inadequacy of the expert's argumentation by requesting clarifications. "We don't really see what the problem is"; "What does this have to do with safety?"; "What is the definition of good training?" Such reactions were observed during the discussions on the report on training course cancellations (skills management assessment). The likelihood that the licensee will make such challenges is particularly dreaded within the SEFH. It seems to be internalized by the department's proofreaders and the head of the SEFH in particular. In the case of the Minotaure review assessment, the latter asked the expert several times to justify the ties to safety and show how the gap with the reference

---

[258] Article 1 of French decree No. 2002-254 of February 22, 2002, relating to the IRSN.

model poses a hazard for the facility. Basically, he asked the expert to use cognitively strong arguments in order to win over the licensee.

Furthermore, by making it possible to strengthen the cognitive links between nuclear safety and human and organizational variables, the causal analysis is also a first step toward enhancing knowledge in the human factors assessment. In order to effectively improve the assessment, the knowledge acquired during it requires additional work. Once the assessment is finished, this knowledge must be capitalized and feedback shared.

Although the causal analysis seems to make it possible to "maximize" rhetorical and cognitive effectiveness, it is nevertheless a costly stage due to the skills and amount of collaboration it requires. For example, during a safety review, the expert must select specific hazardous activities with other experts and may even have to construct the causal chains likely to lead to a hazard with the licensee. Consequently, implementing such a strategy can jeopardize meeting the assessment's time constraints.

3.4. Conclusion: Persuade or Convince?

Chaïm Perelman's 1958 treatise, which returned credibility to rhetoric and was written with Lucie Olbrechts-Tyteca, gives various distinctions between the verbs "persuade" and "convince". These distinctions may appear to form a historical boundary between advocates of truth and advocates of opinion, "between philosophers seeking the absolute and rhetors involved in action"[259]. Thus, "to the person concerned with results, persuading surpasses convincing, since conviction is merely the first stage in progression toward action. Rousseau considered it useless to convince a child 'if you cannot persuade him'." On the other hand, "to someone concerned with the rational character of adherence to an argument, convincing is more crucial than persuading .... In Pascal's view, persuasion is something applied to the automation – by which he means the body, imagination, and feeling, all, in fact, that is not reason."[260]

Perelman is not satisfied with these boundaries. To him, they are based on a selection of abilities and reasonings in the name of a rationality that he considers conventional. He illustrates the failings of such a process with a very simple example. "For example, one will tell us that a person, convinced of the danger of chewing too quickly, will nevertheless continue doing so; The reasoning on which this conviction is entirely based is isolated. One forgets, for

---

[259] Perelman, C. and L. Olbrechts-Tyteca (1958 (2000)). *Traité de l'argumentation*. Bruxelles, Editions de l'Université de Bruxelles, p. 35.

[260] Ibid., p. 35.

example, that this conviction may run counter to another conviction, the one that tells us that eating faster saves time. One therefore sees that the conception of what constitutes conviction, which may appear to be based on a differentiation of the means of proof or abilities at play, is often also on the isolation of certain types of data within a much more complex set."[261]

Nevertheless recognizing the existence of a "definable nuance" between the two notions, Perelman proposes a distinction predicated on another criterion, which he describes as being "quite similar in its consequences yet different in its principle" in comparison to that proposed by Kant. "We are going to apply the term *persuasive* to argumentation that only claims validity for a particular audience, and the term *convincing* to argumentation that presumes to gain the adherence of every rational being."[262]

We will use this distinction to characterize the two strategies analyzed. In a way, to use the reference model to justify a recommendation is to want to persuade a particular audience, one made aware of human factors and the benefits of the reference model, or possibly a licensee aware that it has not provided sufficient documentation. This audience exists, for we have seen it; recommendations justified solely by gaps with the reference model are sometimes accepted (cf. the Minotaure review and skills management assessments). By using cognitively strong arguments that justify the causal link with safety, the expert aims to use a convincing discourse, "one whose premises are universalizable, that is, acceptable in principle to all the members of the universal audience"[263]. In short, a discourse likely to persuade the most doubtful licensee.

We have thus explained two standard solutions to the central problem of the assessment. The first falls within an institutional rationality. It makes overcoming demanding constraints possible and requires less analytical effort. It can lead to persuasive recommendations, but does not make it possible to enhance the knowledge of the human factors assessment. The second falls within a cognitive rationality. It can lead to convincing recommendations that make it possible to achieve a high level of rhetorical and cognitive effectiveness. However, this solution is expensive to implement and can lead to missed deadlines. The importance of meeting deadlines is a constraint that we have emphasized.

---

[261] Ibid., p. 36.

[262] Ibid., p. 36. This definition of conviction led Perelman to propose the concept of "universal audience". (cf. Schmetz, R. (2000). *L'argumentation selon Perelman: pour une raison au coeur de la rhétorique*. Namur, Presses universitaires de Namur.)

[263] Perelman, C. (1977 (2002)). *L'empire rhétorique. Rhétorique et argumentation*. Paris, Librairie philosophie J. Vrin, p. 37.

The results of our analysis are summarized in Table 18. By incorporating the types of effectiveness of their activity, our model may better guide the experts in making choices. However, it does not answer the question of what are the assessment's potential effects on the facility.

	RATIONALITY	
	INSTITUTIONAL	**COGNITIVE**
Chief objective	Persuasive recommendations	Convincing recommendations
Standard analysis	Comparison with the reference model	Causal analysis
Potential rhetorical effectiveness	Satisfactory	High
Potential cognitive effectiveness	Low	High (sharing necessary)
Effort to understand technical processes	Not necessary	Indispensable
Cost of implementing the standard analysis	Low	High (demonstration of safety, collaboration with the technique's experts)
Potential bounding constraints	Incompatibility with the incident analysis	Time constraint, difficulty for the general themes

Table 18: Institutional rationality versus cognitive rationality

Chapter 7. The Operational Effectiveness of the Assessment — Mastering the Strengths of Technical Dialogue

A discussion of the operational effectiveness of the assessment can be approached in various ways.

The production process that is the assessment, whose operations are described in the preceding chapter of this book, result in a final outcome— a list of conclusions and recommendations. The following question immediately comes to mind: what effects do these recommendations have on the safety of the assessed facilities? This question imposes the choice of a conventional definition of safety – safety learned following a number of incidents, safety as a set of barriers, safety as a dynamic non-event protected by socio-cultural processes. Such a (possibly hybrid) choice would raise a number of tricky questions – How can a reduction in the number of incidents be attributed to the application of a recommendation? Do the barriers in place make it possible to effectively avert an incident? How can the socio-cultural processes that would enable the incident to be averted be characterized? To answer these questions and be able to hope to establish a link between the experts' recommendations and safety would require having a sufficient number of skills and resources and involve active participation of the facilities' employees and the experts' central services.

Our place in the Department for the Study of Human Factors was not well suited to such a project. Also, the project's duration would not have fit with our schedule commitments. In addition to the recommendation implementation periods, a lengthy observation period would have been necessary in order to identify their effects. Furthermore, although the researcher will find such a conception of the operational effectiveness to be certainly relevant and likely to clarify the links between the organizational factors and nuclear safety, the prospect of not being able to issue an opinion on the operational effectiveness of an assessment until several years after the licensee receives the recommendations is awkward.

So then, how did we evaluate *ex post* the operational effectiveness of an assessment without having to scrutinize the tangible effects on the facilities? To begin with, we chose to analyze the recommendations in order to identify the potential effects of the human factors assessment. As we present the formal control methods designed to ensure the materialization of these effects, we will underscore their limitations and predict a low operational effectiveness. However, limiting the assessment to its recommendations and the inspection to the formal post-assessment process ignores the main vector of the assessment – technical dialogue. Listing the effects[264] of the human factors assessment and identifying the systems that encourage the licensee to implement the recommendations requires taking into account the sequence of the assessment process. It is through this sequence that a continuous technical dialogue takes place between the stakeholders of the human factors assessment, which is a true source of its operational effectiveness. In addition to the skills that they must master in order to attain satisfactory levels of rhetorical and cognitive effectiveness, other types of knowledge are necessary to make their assessments effective.

1. FROM RECOMMENDATION TO ACTION – THE POTENTIAL EFFECTS OF THE ASSESSMENT

We approach the effects of the human factors assessment through its recommendations. These effects remain assumed; indeed, the recommendations do not appear to be routinely implemented. Furthermore, such implementation requires an in-depth evaluation that does not seem to be part of the formal post-assessment process. Without a suitable incentive system, the operational effectiveness appears to be low. However, these conceptions of the human factors assessment (via the recommendations) and the post-assessment process (reduced to the formal process) are insufficient to properly understand the operational effectiveness of the human factors assessment.

1.1. A Typology of the Expected Effects

Using the results of the analyses conducted in the previous chapter leads us to distinguish two standard effects – compliance effects and learning effects. Our interviews with the experts' contacts following the assessments led us to add a third type of effect – legitimation effects.

---

[264] The identified effects stem from the data reconstructed previously as well as from interviews conducted with the human factors specialists after each of the three assessments.

Compliance and learning effects

The analyses of the recommendations of five human factors assessments led us to identify a reference model characterized by human and organizational factors that we have proposed categorizing under the five headings of operating experience feedback system, human-machine interfaces, management of the documentation system, skills management process, and organization of work. We have seen that the vast majority of the experts' recommendations suggest assigning the reference model's value to the human and organizational variables of the facilities assessed. Implementing these recommendations should thus result in bringing the said facilities further into compliance with this model. The potential effects of the human factors assessment are thus primarily effects of compliance with the reference model.

By indicating gaps with a reference model, these recommendations also may allow the licensee to ponder the configurations of its organization. Furthermore, we distinguished two recommendations (cf. Table 9, p. 222) in the Minotaure review assessment. These recommendations do not bring the facility in line with the reference model but require the licensee to analyze its hazardous activities, explain the related risks and justify the preventive measures. Since implementing these recommendations requires conducting causal analyses, it should allow the licensee to enhance the state of knowledge on human and organizational factors. As a result, the recommendations also produce learning effects.

Legitimation effects

During our interviews of two CEA representatives who had participated in the Minotaure review assessment, we noted that the human factors specialists' recommendations could also help their approach and make their work easier. We must point out that although talk about the importance of human and organizational factors is commonplace and although the ASN has an ambitious action plan, the discipline is sometimes barely represented within the facilities and in design projects. "The commitment to take human factors into consideration in the incident analysis goes in the right direction. We're of course for it and the SEFH helps us with it. We're the first ones to promote analyses of sensitive activities. So, it's a help." (interview of July 13, 2006) "In my view, reviewing the formalism of the modification request to incorporate risk analysis goes in the right direction because researchers usually send us modification requests without any risk analyses. I also agree regarding foreseeable and unforeseeable associated operations. I'm not against strengthening the safety function means." (interview of July 7, 2006)

Thus, among the expected effects of the assessment produced by the recommendations, we also identified collateral effects, or legitimation effects.

The recommendations legitimized the behavior of certain workers at the facilities, in particular those who design and implement safety enhancement measures.

1.2. Difficulties of Control

Once the assessment is finished, the recommendations of the human factors expert are changed to commitments[265] taken by the licensee and requests of the ASN. Consequently, the licensee must send its replies and implement the recommendations within a time period stated in a summary letter sent by the ASN. However, as the recommendations are liable to run counter to some of the licensee's objectives, they are not systematically implemented. Furthermore, the control methods implemented in the formal post-assessment process led by the ASN do not seem to be fully suited to the recommendations of the human factors specialists.

In the Artémis incidents assessment, the experts considered some preparation and activity inspection methods to be insufficient. Furthermore, the "interfaces" between the researchers – assessed primarily on the basis of research objectives – and the licensees – responsible for the safety of the facility – were seen by the experts, the ASN's representatives and several of the facilities operators as the main problem.

Researchers are anxious to have their research results in time, and some safety measures can slow down their work. More generally, following certain safety principles can sometimes be a constraint for workers who are evaluated on the operational results of their facility[266]. Consequently, it can be feared that some recommendations are not scrupulously implemented, particularly since they often require resources and the licensee often has limited means. "I don't know how to do an ergonomic workplace assessment. At least a half-day is necessary per shielded process line. The problem is the number of hours of

---

[265] This term refers both to the CEA's commitments and EDF's stands/actions.

[266] This phenomenon is not specific to facilities. In his book on industrial accidents, Michel Llory writes of "production pressures" caused by the necessity to achieve objectives and which are likely to lead to risky behaviors (cf. Llory, M. (2000). *Accidents industriels: le coût du silence. Opérateurs privés de parole et cadres introuvables.* Paris, L'Harmattan.) In a critical analysis of what he considers to be the dominant safety paradigm, Claude Gilbert writes that "the safety of people is one priority among others for organizations that run hazardous activities. Performance, competitiveness, the continuation of operations, the preservation of capabilities of technological innovation, experimentation and the such are other imperatives for these organizations that are faced with market or government constraints." (91) (cf. Gilbert, C. (2005). Erreurs, défaillances, vulnérabilités: vers de nouvelles conceptions de la sécurité? *Risques, crises et incertitudes: pour une analyse critique.* O. Borraz, C. Gilbert and P.-B. Joly. Grenoble, Publications de la MSH-Alpes: 69–115 (257).)

work compared with the means available. Training people in the modification procedure takes work, but it's not a bad idea …. However, it'll be done to the detriment of other things …." (Artémis, the licensee, interview of July 7, 2006)

Of course, the licensees seemed to take seriously the fact that these recommendations were changed to commitments or requests from the ASN. "We've made a commitment to formalize the actions and to be accountable for it." (Minotaure, the licensee, interview of July 13, 2006) "It's not going to be easy to meet the deadlines, but there are no two ways about it." (Artémis, the licensee interview of July 7, 2006)

Furthermore, it is not unreasonable to think that implementing a recommendation should be made easier when a licensee is persuaded of its relevance and committed to it. "Committing to the requests wasn't a problem for us because I think that we would have carried them out anyway." (Minotaure, the licensee, interview of July 13, 2006)

But such is not always the case, as illustrated by the following reaction of the experts when they read the licensee's responses to the recommendations they had drafted during the operating organization assessment (cf. Table 13, p. 222): "We asked them to make analyses and what do we get? A single page of answers!" (the head of the SEFH, informal discussion of March 13, 2007)

It is therefore clear that the recommendations are not systematically followed. This problem is experienced much more by the ASN (regulatory body) than the IRSN (assessment institute). Engineers in the ASN's central services are in charge of following the various cases and verifying that licensees provide responses. Moreover, for several years now the engineers in the ASN's regional divisions have been conducting one-off inspections that focus on human factors.

Such measures may be sufficient to elicit a response from the licensee. But the question is, what kind of response? The case of the operating organization assessment prompts caution, for such responses can be very succinct. It appears that adequate follow-up requires making an in-depth evaluation of the licensee's response. Indeed, simply verifying the sending of a response might have "overlooked" the response given by the licensee in the case of the operating organization assessment. The actions taken to "draw up a report on the effects of the reorganization", "improve the consideration of human factors in the incident analyses", "explain a method for identifying skills needs", "perform an ergonomic analysis of all workplaces", or even "draw up a training postponement report" call for close examination that cannot be limited to procedural control or result-based control.

Although the inspections can be of use to the human factors specialists – we saw in the case of skills management that the experts used the findings of the ASN engineers to frame the review – its control methods do not appear to be wholly suited to the evaluation of the human factors assessment recommendations.

1.3. Conclusion: The Limits of Analysis

Based on this short analysis, the operational effectiveness is understood to mean two things. Firstly, some assessment recommendations are likely to interfere with certain interests within the facilities, in which case the licensee may not implement them to the letter and their potential effects (compliance, learning and legitimation) are not ensured. Secondly, the formal post-assessment process, limited to ASN inspections, does not routinely integrate an in-depth evaluation of the licensee's responses. This process therefore would hardly incite the licensee to give quality responses. This leads us to predict a potentially low operational effectiveness.

However, this analysis is based on a relatively poor conception of the assessment, to wit, a list of recommendations initially devised by experts who would be absent from the post-assessment process. Nevertheless, the data described in the previous part show that this conception was unsuited. Indeed, 1) we have seen that the human factors assessment was not to be viewed merely as a set of recommendations, but rather as a process of interactions likely to produce effects to be factored into our evaluation of the operational effectiveness. 2) We have also seen that the assessment of incidents at the Artémis facility was followed by an assessment of safety at Artémis and in which the experts participated. As a result, the experts can take part in the post-assessment process, which should not be viewed merely through the lens of ASN inspections.

A broader conception of the assessment – an assessment-process that is one element in a series of assessments – is thus necessary to evaluate the operational effectiveness. Taking such properties of the human factors assessment into consideration may lead us to reconsider our relatively negative initial prediction.

2. REGULATION THROUGH TECHNICAL DIALOGUE

Viewing the assessment not just as a list of recommendations associated with a simple form of control, but rather as an assessment-process that is one element in series of assessments. Such an approach takes into consideration the particularities of nuclear safety regulation in France – a *technical dialogue* maintained by the stakeholders within the framework of continual assessment not bounded by time. Using the various cases discussed herein, we intend to identify the effects attributable to this technical dialogue.

Maintaining a dialogue with the licensee is the incentive mechanism that seemed to be lacking from the formal post-assessment process. The assessment itself ($n + 1$) entails making a detailed evaluation of the licensee's responses to the requests stemming from the assessment (n). Furthermore, in the previous part we identified interactions that, within the course of the assessment process, might produce tangible effects on the facility. Such effects must be taken into consideration when assessing the operational effectiveness of the assessment. Because it prompts the licensee to respond to the aforementioned requests and produces effects during the assessment process, technical dialogue is an important vector of the operational effectiveness of the human factors assessment.

2.1. The Effects of Continual Dialogue

The two successive assessments at the Artémis facility illustrate the relevance of seeing the assessment as an element of a sequence. Such a view makes a detailed evaluation of the licensee's responses by the SEFH experts – the necessity of which we have underscored – possible. Accordingly, several responses provided by the licensee after the Artémis incidents assessment were evaluated during an assessment of the review of safety at Artémis. For example, the working conditions of the operations managers and the coordination arrangements between researchers and safety engineers were re-evaluated.

Likewise, at that very moment, the human factors expert who had participated in the Minotaure review evaluated, as part of a new assessment, the licensee's response regarding the analysis of the fuel rod fabrication activity. He did so by collaborating with the generalist engineer in charge of coordinating the facility review and performed a causal analysis.

As for the skills management assessment, it was perhaps just the first in a series of reviews exclusively aimed at the theme, as suggested by the following comment from an ASN representative: "It was the first advisory committee meeting on skills management. We now see how the skills management system works and can say whether EDF's processes are good. This first meeting is on target." (interview of August 23, 2006)

If a second skills management assessment is conducted, part of the assessment will most likely focus on how the requests and commitments stemming from the first report were dealt with. Furthermore, the licensee's responses regarding management of the skills of the monitoring officers were evaluated during an assessment that focused on maintenance and was presented to the advisory committee in 2008.

Assessments were thus primarily used to evaluate the implementation of the human factors specialists' recommendations. The continual dialogue between the experts and the licensee, maintained by this series of assessments, contributes to make real the potential effects of the recommendations. It is interesting to note that what seems to have a positive impact on the operational effectiveness of the assessment is in actual fact a property that runs counter to traditional conceptions. Just like the judicial expert assessment, which one can hardly imagine continuing after process ends, the canonical model of expert assessment is sporadic, time bound, whereas the nuclear safety assessment seems to be in a perpetual state of incompletion.

In addition to the effects produced by the recommendations and ensured by the continuity of technical dialogue, there are also the effects created by the interactions that make up the assessment process.

2.2. The Effects of Deterrence and Compliance

The technical dialogue that takes place between the regulators and the licensee throughout the assessment process is a central mechanism of French safety regulations. In order to best describe it, we will use a distinction that turns on the forms of regulation. Used by several researchers in the Anglosphere, this distinction is based on the type of relationship between regulators and those regulated – compliance and deterrence. We endeavored to describe the nature of the regulator-licensee interactions and identify their outcomes for each of the three assessments described in this book. Some were only intermediate outcomes that became recommendations during the assessment process. Others were tangible effects on the facility that were distinct from the recommendations and must be taken into consideration to evaluate the operational effectiveness of the assessment. We will present these effects in a summary table and describe the interactions that created them and give the value taken by the "form of regulation" variable (compliance/ deterrence), for technical dialogue integrates forms of regulation through both compliance and deterrence.

Compliance versus deterrence

In their book on risk regulation regimes, Hood, Rothstein and Baldwin write that "one of the most prominent debates over behavior modifications [of those regulated] in the law and regulation literature concerns the relative merits of compliance and deterrence as ways of applying legal or regulatory standards."[267] The chosen form of regulation is one of the variables used by the authors to describe several risk regulation regimes.

---

[267] Hood, C., H. Rothstein, et al. (2004). *The government of risk. Understanding risk regulation regimes.* Oxford, Oxford University Press, p. 27.

Compliance and deterrence are distinguished especially by the nature of the relationships between the regulator and the regulated: "'Compliance' doctrines rely heavily on diplomacy, persuasion or education rather than routine application of sanctions to produce a compliance culture on the part of those affected by the regulation. By contrast, 'deterrence' doctrines, going back to Bentham and Beccaria, rely on the credibility of penalties or punishment, expressed in the 'expected cost' of non-compliance to violators, to prevent those regulated from breaking the rules."[268]

In our historical review, we described France's nuclear safety institutions as implementing a technical dialogue that replaced in particular a set of legal provisions and financial penalties. Consequently, there is reason to think that a flexible form of regulation is prevalent. However, as we will see, some interactions between the regulator and the regulated come under a form of deterrence.

Analysis of the effects

During the Minotaure review assessment, the licensee presented several documents proving that human factors had been taken into consideration in the control room renovation project. In the view of one CEA human factors expert, this integration of human factors right from the design stage was a significant step forward. "A big improvement was made at Minotaure in that we were contacted right from the design stage. You must bear in mind that there was a real motivation − particularly on the part of the renovation project leader − to integrate human factors in the execution of the project. Furthermore, an investigation was conducted." (interview of July 13, 2006)

The project leader's motivation was prompted in particular by a letter, sent by the ASN before the assessment, that repeated a request from the IRSN demanding that special efforts be made regarding human factors. "The ASN sent a letter explicitly asking us to study human factors. Of course it had an effect!" (Minotaure, the licensee, interview of July 13, 2006)

A second effect bears mentioning. Recall that a warning letter[269], cosigned by two IRSN department heads and sent during the assessment, prompted the licensee to send an in-depth analysis of the fuel rod fabrication activity that listed the human and organizational arrangements for preventing the criticality

---

[268] Ibid., p. 27.

[269]Such a warning could lead to a limitation of reactor operation: "... we draw your attention to the fact that whenever it is not demonstrated that measures meet FSR I.3.c, it is customary to set authorized mass limits for fissile materials that are well below the acceptable limits. Such reduced limits case could severely constrain facility operation."

risks of these sensitive activities. The letter had the effect of prompting the licensee to conduct a causal analysis that, as we have said, contributed to enhancing knowledge.

During the assessment relating to the two incidents at Artémis, we did not identify any actions by the licensee that might have been directly prompted by the involvement of the experts or the ASN's representatives However, it is reasonable to think that the presentation – given by the facility manager at the request of the expert and the IRSN coordinator – of the corrective actions and their deadlines created additional "pressure" to implement the action plan and thus establish several human and organizational factors.

During the framework stage of the skills management assessment, the human factors specialists succeeded in convincing EDF's coordinators of the relevance of their assessment approach (occupation studies conducted at several sites). It can be assumed the reports they routinely made to each site's representatives were a source of learning for the latter.

It seems that, during the review, several of the experts' requests led to the drafting of summary documents. "Many of the documents provided date from the start of the review." (the coordinator, IRSN internal meeting of December 21, 2005). One of the IRSN representatives saw this as a positive effect of the assessment. "This is the first effect of the assessment." (IRSN internal meeting of December 21, 2005)

Also recall that a summary document was drafted after an ASN representative spoke up at the halfway meeting. By clarifying and formalizing the different skills management policies and how these documents are created, the licensee complies further with the experts' human factors reference model.

Lastly, the licensee's many alignments, identified during the feedback meetings and the meetings held in preparation of the advisory committee meeting (cf. p. 204), can be viewed in part as learning effects. This is what EDF's representatives appeared to say at the end of the assessment, in particular regarding the systems for managing monitoring skills. "You showed us a few things about monitoring. We thought that we knew more than you did." (post-assessment feedback meeting of June 16, 2006). An occupation coordinator's highly positive opinion of the reading of the study report on "his" occupation can also be mentioned. "A lot of people were surprised by the strength of the analysis and conclusions. I find your report to be of very good quality and I use it." (conference call of October 24, 2005)

Table 19 summarizes the effects that we have just discussed and shows a few of their characteristics.

Case	Event (stage)	Interactions	Form of regulation	Effects
MINOTAURE REVIEW	Request to take human factors into consideration right from the design stage (prior stage)	ASN – CEA	**Deterrence** (letter from the ASN)	Compliance with the reference model, legitimation of the CEA's experts, learning of human factors by the licensee
	Request for clarification on the arrangements regarding the criticality risk (review)	IRSN – Minotaure	**Deterrence** (threat of limiting operation)	Learning (hazardous event causal chains)
ARTÉMIS INCIDENTS	Presentation of the action plan by the facility manager and safety engineer (review)	Human factors expert and coordinator-Artémis	**Compliance** (discussions between the coordinator and the licensee)	Compliance with the reference model
SKILLS MANAGEMENT	Request to access actual facilities followed by reporting of the occupation studies (scope definition)	HF experts – EDF coordinators	**Compliance** (presentation by the experts of the assessment's objectives)	Learning (reporting of the occupation studies to the sites and central services)
	Documentation request (review)	HF experts – EDF coordinators	**Compliance**	Compliance with the reference model
	Request for a summary (review)	ASN – EDF coordinator	**Deterrence** (taking of the floor by the ASN)	Compliance with the reference model
	Reports of interim versions of the report (review)	HF experts – EDF coordinators and coordinators	**Compliance** (open discussions)	Learning (alignment of the licensee during the preparatory meeting)

Table 19: The effects of compliance and deterrence observed during the assessments followed.

The table above shows in particular that the forms of regulation used can fall under the rubrics of compliance as well as deterrence. Technical dialogue is therefore a hybrid approach. Some authors advocate taking such an approach. For example, Ayres and Braithwaite (1992) are in favor of implementing a flexible form of regulation when the regulated lacks sufficient knowledge or demonstrates a desire to make progress. On the other hand, a deterrent approach must be adopted when the regulated are opportunistic or immoral.

The table also shows that the assessment's effects are created not just by the human factors specialists but also by their generalist engineer colleagues (coordinators and management for the Artémis incidents and Minotaure review assessments) and the ASN's representatives (Minotaure review and skills management). The collaborative aspect of the assessment thus contributes to its operational effectiveness.

2.3. Conclusion: The Strengths of Technical Dialogue

Technical dialogue is thus what brings about the significant effects of human factors assessments. Its strengths come in particular from the many forms of relationships it allows with the licensee – by involving the experts, generalist engineers, the institute's management and the ASN's representatives – during an assessment process that seems to be virtually unlimited in time.

It seems that the technical dialogue's continuity makes it possible to overcome potential changes in people, as illustrated by the following statement of a representative of the Minotaure facility: "The IRSN's requests are also a message within the IRSN, for when there are different experts ... ." (interview of July 13, 2006) During future assessments, technical dialogue also makes it possible to verify compliance with recommendations or rather to evaluate the quality of the human and organizational factors added following the requests and commitments. It also makes it possible to compensate for certain deficiencies in the recommendations caused by the imprecise values assigned to the reference model's organizational variables: "We've been asked to do an in-depth analysis of the operations activity, but we haven't been told much else .... We'll propose something and then technical dialogue will take its course." (a CEA expert, interview of July 13, 2006)

Although technical dialogue does not preclude forms of regulation by deterrence, the effects produced by flexible forms of regulation based on compliance – particularly learning effects[270] that benefit the licensee and the experts – must be emphasized. The comments of the leader of the control room renovation project at Minotaure and EDF's skills management coordinators on the quality of the assessment reports can be mentioned in this respect. "*In short, your report is good.*" (skills management, post-assessment feedback meeting of 16 June 2006). "The IRSN has done an outstanding job. The advisory

---

[270] Franck Aggeri also highlighted this in the field of environmental policy, using the case of vehicle recycling (cf. Aggeri, F. (1999). "Environmental policies and innovation. A knowledge-based perspective on cooperative approaches." *Research policy*(28): 699–717.)

committee's report is a summary with lots of information. We use it. It's a reference on the facility." (Minotaure, the licensee, interview of July 13, 2006)

There is thus every reason to believe that these reports will be consulted and used. The Minotaure review assessment report thus might make it possible to raise further awareness on hazards at the facility, while the skills management assessment report might help decision-makers in the central services and local skills management officers. "I find the IRSN report very well done and refer to it a lot. It summarizes everything that is done centrally." (a skills management officer at a nuclear power plant, interview of December 19, 2006)

Several of the effects of the technical dialogue were not directly produced by the experts during their assessments. Furthermore, we earlier explained the tough objectives that they endeavored to meet while working within tight constraints. So we can legitimately ask ourselves how they could (further) integrate operational effectiveness objectives into their activities. Could they better control the strengths of technical dialogue during the assessment process?

3. CONTROLLING TECHNICAL DIALOGUE

In the previous chapter we highlighted several types of knowledge the experts had to master in order to stay within the institutional constraints of the assessment, meet alignment objectives and enhance knowledge. Putting such knowledge into practice can nonetheless turn out to be insufficient to make the assessment operational. Accordingly, the experts must do as we have – consider their contribution as an element of a continual sequence. Furthermore, some regulator-licensee interactions during the assessment process may have effects on the licensee's safety practices or representations. It is therefore important that the experts know how to "trigger" and control these particular interactions. Once the skills required to control technical dialogue are set out, the entire range of knowledge the expert must possess in order to conduct a rhetorically, cognitively and operationally effective assessment can be summarized.

3.1. Managing Continuity

Viewing the assessment as a link in a chain has significant consequences on the expert's work. Whether the assessment focuses on a theme (skills management) or a facility (Artémis, Minotaure), the expert's work must first have a historical dimension. In order to bolster the relevance of his argumentations and facilitate his assignment, the expert must become acquainted with the work performed in the department (assessments of training for the skills management case, analysis of incidents for the Artémis review contribution) and take into account the safety record of the facility or facilities (incident reports in particular). In order to enhance the operational effectiveness of the assessment, the expert must also assess the responses to the human factors requests made during the previous assessment in the chain.

A "handover" may sometimes take place within the department. This occurred during the two Artémis assessments. During the scope definition stage, the contributor to the safety review received the assistance of the incident analyst. The analyst was familiar with the facility and had identified themes to be explored more closely[271].

A future-forward direction must be taken when formulating recommendations. The specialist (n) must think about facilitating both the implementation of these recommendations by the licensee and the follow-up work of the ASN's engineers and of his colleagues. His recommendations must be sufficiently clear so that, simply by reading them, the specialist ($n + 1$) can assess how to implement them and understand the safety issues.

3.2. Managing Interactions

The specialist is directly involved in many of the regulator-licensee interactions in the assessments that we followed. In many instances the data of the assessment are at work in these interactions. Although these interactions can have tangible and virtually instantaneous effects on a facility, they first must make it possible to acquire data that supplements the data provided by the licensee. An acquisition method similar to the "intervention model", used by management researchers and ergonomists in particular, is necessary in order to address certain issues. In the context of the assessment, however, it may create capture phenomena that must be controlled.

Collect or build data?

Data are required to review any type of dossier, regardless of its characteristics. Generally, data provided by the licensee are insufficient. The first reason for this is that the specialist may find them to be incomplete (lack of a chapter on human factors in the operating experience feedback file for the Minotaure review case, lack of data on the preparatory phase for the Artémis incidents case, lack of summary documents for the skills management case). The other reason is that the specialist prefers obtaining additional data rather than placing blind faith in documents that may be viewed as secondary sources (e.g. human factors investigation, incident reports, managerial policies).

---

[271] We participated in this stage as a human factors specialist.

In order to address certain issues, this data must primarily make it possible to use the reference model and reconstruct incident scenarios (either actual, as in the Artémis incidents assessment, or potential, as in the Minotaure review assessment). The interactions with the licensee consist of data collection. They can be seen as a transmission from the licensee (which may be represented by a facility manager or operators) to the expert. The underlying "knowledge production model" is similar to what Armand Hatchuel suggests calling the "field model" whereby the researcher (here, the expert) naturalizes an object (the organizational factors of the reference model, the event chains) in order to study it[272].

The knowledge production model primarily used in the case of skills management is markedly different. The necessary data are distinct from the data in the previous cases in that they are not based on any knowledge possessed by the representatives of the facility's operation. It involves having information on how the operators and operating crews perceive the skills management systems and the practical use of the management tools. Admittedly, such data are "collected", but they are subsequently presented to the site representatives, discussed with EDF's coordinators and the occupation coordinators, and compared with the policies developed in the central services. We have also seen in this case that the field had to be defined with the help of EDF's coordinators (choice of occupations and sites). It is these reasons that bring the knowledge production model closer to the "intervention model"[273].

The data obtained and the analyses performed may reveal courses of action in use at the licensee's and thus allow the decision-makers in the central services and the site representatives to think up new management methods. Consequently, the specialists' work may create learning (cf. Table 19). It is these learning effects that often justify the value of management intervention research.

---

[272] Hatchuel, A. (2001). "The two pillars of new management research." *British journal of management* **12**: 33–39.

[273] Moisdon, J.-C. (1984). "Recherche en gestion et intervention." *Revue française de gestion.* 61–73, Engel, F., D. Fixari, et al. (1986). Le séminaire "Pratiques d'intervention des chercheurs". *Chercheurs dans l'entreprise ou la recherche en action.* C. Alezra, F. Engel, D. Fixari and J.-C. Moisdon. Paris, Les cahiers du programme mobilisateur "Technologie, Emploi, Travail". Ministère de la recherche et de la technologie. **2**: 11–35, Hatchuel, A. (1994a). "Les savoirs de l'intervention en entreprise." *Entreprises et Histoire*(7): 59–75, Guérin, F., A. Laville, et al. (1997). *Comprendre le travail pour le transformer. La pratique de l'ergonomie.* Lyon, Editions de l'ANACT, Rolina, G. and L. Roux (2006). La recherche-intervention en gestion. Une généalogie de l'intervention au CGS. *Intervenir dans le monde du travail: la responsabilité sociale d'un centre de recherche en sciences humaines.* Liège, Education_permanente (2007). "Intervention et savoirs: la pensée au travail." *Education permanente*(170).

Although this method of knowledge production – which seems essential for addressing certain managerial objects – thus may considerably influence the operational effectiveness of the assessment, it can place the expert in a tricky position. Indeed, the intervention model is defined by a certain type of relationship between the researcher and the organization that often prompts the request for intervention. Its use in the institutional context of the assessment can be tricky.

Controlling capture

In each of the three cases we have described in this book, negotiations were necessary in order to access actual facilities. This was especially true for skills management, where the field had to be defined with the representatives of the central services. This indispensable collaboration led the specialists to accept compromises, narrow down the themes of the assessment and revise the list of occupations selected for their studies.

As we have already pointed out, this restriction of the specialists' freedom imposed by the licensee can be interpreted as a capture phenomenon. It was particularly felt by the specialists during a feedback meeting. They felt that they were being treated like internal auditors or consultants. In short, they were contractors working for EDF's coordinators. A post-assessment feedback meeting held on June 16, 2006, provided the experts and EDF's coordinators the opportunity to discuss the matter:

– EDF: One of your teams hadn't earned trust or legitimacy. We had to contend with ill-tempered sites that didn't want to participate in the study.

– EDF: We devoted time to you, but the people at Saint Guillaume didn't want to see you. We put on the *pressure*. We expect something in return, which means you have to be relevant.

– The head of the SEFH: Yes, but we work neither for you nor directly for the ASN. We work for an abstract entity known as safety.

As Ayres and Braithwaite (1992) and La Porte and Thomas (1995) have noted, capture of the regulator can be good for safety. The data that the momentarily "captured" specialists were able to obtain produced learning effects, likely to enhance safety, within the licensee's facility. The following comments of an EDF representative illustrate the positive aspect of capture: "Likewise, we're using you to get things going! We started off with the idea that you were going to teach us things, which is the case because you've taught us things about the monitoring officer. That was our basis for choosing the sites. We didn't show you the NGD's showcase! We took risks! It's management by the ASN [and the IRSN]. We need a safety authority that pushes us on certain matters." (Skills management, post-assessment feedback meeting of June 16, 2006)

However, there is reason to think that it is necessary that this capture be only momentary and that the expert succeed in breaking free of this capture in order to regain a certain independence, particularly when formulating his conclusions and decisions. In a way, the head of the SEFH indicated as much during the aforementioned meeting when he told EDF that "we don't have to completely abide by your relevance criteria. The expert must have a different perspective." Our data have shown this independence of the expert during the drafting of the assessment report. The meetings held in preparation of the advisory committee meeting have shown that the specialists did not bow down to the licensee's "relevance criteria". Furthermore, in view of the continuity of the assessment, it can be said that negotiations on the wording of the requests, when such requests pertain only to deadlines, result only in minor changes. Controlling capture thus depends greatly on the clear stands that the specialists are able to take when face to face with the licensee.

Quality documentation sent right at the start of the assessment might also make it possible to better clarify roles:

– The SEFH coordinator: There was no dossier [for the advisory committee meeting on skills]. Instead, documents were sent in haphazardly. There wasn't a summary to explain how everything was tied together ...

– A representative of the DSR: ... which led us to do their job for them, I suppose?

– The head of the SEFH: EDF needs to provide a real dossier. Otherwise, we're going to spend our time doing it for them and then – and I'm overstating things a little – they'll the ones who'll assess the quality of our dossier! (Skills management, feedback meeting, April 13, 2006)

We have seen that the regulation by deterrence approach adopted by the ASN's representatives and the IRSN's management was likely to get the licensee to provide such documents – an investigation and analysis of the fuel rod fabrication activity at the Minotaure reactor and a (long) overdue summary document for skills management. Nevertheless, within nuclear safety institutions, it is not unreasonable to think that this form of regulation must be used sparingly so as not to jeopardize the quality of a continual dialogue built to ensure technical discussions.

3.3. Conclusions: The Expert's Skills

The accounts in the second part and all our analyses have highlighted the fine points of the assessment activity and the many skills that the human factors specialists must master in order to achieve satisfactory levels of rhetorical, cognitive and operational effectiveness. These skills are listed in Sidebar 24.

Based on their study of several expert systems, Armand Hatchuel and Benoît Weil (1992) proposed classifying skills into three types: the artisan's skills ("doing" know-how), the repairer's skills ("understanding" know-how) and the strategist's skills ("combining" know-how). The specialists' skills fall under all three types. For example, knowing how to involve management and the ASN's representatives falls under "doing" know-how; knowing how to reconstruct event chains falls under "understanding" know-how; and knowing how to argue and align falls under "combining" know-how.

- *Know when and how to seek the deterrent effect of the IRSN's management or the ASN's representatives to convince the licensee to provide data;*

- *Know how to obtain additional data. To do so, the specialists need to know (1) when and how to interact with the other experts at the institute in order to understand the overall workings of the facility being assessed and identify the risks related to certain activities; (2) how to negotiate with licensee in order to gain access to its facilities; and (3) how to build the field with the licensee and step in while, at the same time, ensuring that they control the capture phenomena related to such collaboration;*

- *Control the two analysis types explained. Knowledge of the reference model is necessary in both cases. Furthermore, basic knowledge of the techniques used at the facility is necessary in order to establish the causal chains;*

- *Know how to factor the facility's safety record into analyses and evaluate the implementation of past recommendations;*

- *Know how to argue, substantiate recommendations designed to overcome internal alignment difficulties (proofreading, approval, integration) with the licensee (comparisons, negotiations) and possibly before the advisory committee;*

- *Meet imposed deadlines and formal constraints and answer the initial questions of the assessment;*

- *Adequately explain recommendations to encourage the licensee to implement them and facilitate the following review, which may be conducted by another specialist;*

- *Compile acquired knowledge and share it with the department's specialists;*

- *Keep up with the scientific literature.*

Sidebar 24: The human factors expert's skills.

By showing that simply mastering a body of established knowledge is not enough to carry out an assessment, this list of skills illustrates the insufficiency of the canonical model in explaining the human factors assessment. Consequently, it questions the traditional organizational forms of assessment activities based on the partnership of autonomous, independent professionals[274].

Indeed, an organization based solely on the autonomy of the experts does not seem sufficient to integrate the continuity of the nuclear safety assessment and acquire and share knowledge. In this sense, organizing specialists by skills areas, making regular situation updates during department meetings and completing feedback forms at the end of assessments are organizational measures that appear necessary for this particular form of assessment. A suggestion could be to round out the systems already in place with leveraging methods that are more formalized and thus less linked to individuals. The creation of a database containing the causal relationships found by the assessments would be useful to the specialists. Statistical analyses could make it possible to identify consistencies, thus strengthening the specialists' knowledge and argumentations, and improve the rhetorical and cognitive effectiveness of the assessment.

That said, the list of skills in Table 24 shows that the SEFH's organization is not the sole factor of its success. Our analyses have shown that the experts had many interaction skills and we have repeatedly mentioned the limitations of the separation of technical factors and human factors in the handling of risks. The integration of the SEFH within the IRSN, and more broadly within the external review system, thus seems to be of utmost importance. However, we have seen some limits to such integration. The absence of the participation of the technique's specialists in the skills management assessment, the vagueness of the request and themes identified during the Minotaure review illustrate the difficulties that the engineers can experience regarding human factors.

A drastic solution would be to break up the SEFH and scatter the human factors specialists among the generalist engineering departments defined by facility type. The post-assessment feedback from the Artémis and Minotaure coordinators and the participation of the technique's experts in in-house seminars that the SEFGH has been organizing for more than a year show the interest experts of the technique can have in human and organizational factors and suggest that the human factors specialists would be welcome within the generalist engineering departments. Such a solution nevertheless seems moot,

---

[274] Gand, S. (2008). L'organisation des dynamiques de services professionnels. Logique de rationalisation, cadre de gestion et formes de collégialité. *Sciences de gestion*. Paris, Mines ParisTech. **Thèse de doctorat**.

for the enhancement of knowledge, which is essential, requires having a group. It would be better to set up a network of human factors correspondents within the generalist engineering departments. Such a move would preserve the SEFH group and could prompt extensive cross-learning between the engineers and the human factors specialists. Generally speaking, any action that might increase the engineers' awareness of the human factors specialists' practices and reasonings could represent a step forward. We hope that our research can contribute to doing so.

Conclusion to Part Three: Restoring the Balance of the Dimensions of Effectiveness

Although our accounts have brought to light the many singular aspects of conducting a human factors assessment, our analyses have allowed us to clarify certain consistencies in the specialists' practices and regulation methods characteristic of the control system that frame their activity. These are the existence of a reference model of organization, the use of two types of analysis (comparison with the reference model and causal analysis), the use of two models of knowledge production (field, involvement) and the practice of two forms of regulation (based on compliance and deterrence, respectively).

After listing the constraints and objectives of the human factors assessment, we proposed a model that explains the specialists' choices. The strength of the constraints and the difficulty of the recommendation and alignment objectives justify the existence of an institutional rationality that prompts the specialist to focus on the rhetorical effectiveness of the assessment, sometimes to the detriment of its cognitive effectiveness, for which there is still considerable scope for improvement.

However, the small amount of knowledge created in the human factors assessment does not prevent it from having tangible effects on the facilities. We have identified three types of effect of this assessment that characterize its operational effectiveness – effects of compliance with the reference model, learning effects and legitimation effects. It is the continuity of a technical dialogue between the regulators and those regulated – one of the specificities of nuclear safety institutions in France – that particularly contributes to the operational dimension of the effectiveness of assessment.

It is particularly for this reason that identifying the tangible effects of the assessment and precisely defining their causes is very difficult. Consequently, suggesting ways to improve operational effectiveness is equally tricky. However, there is reason to think that closer collaboration with the ASN's engineers might improve the monitoring of the assessments and that stating the specialist's recommendations in clear and concise terms might make it easier to implement and monitor them.

On the other hand, the parameters of our explanatory model make it possible to propose ways to improve cognitive effectiveness. First of all, the specialists can be encouraged to use causal analysis. Although researchers have pointed out its challenges and limitations and although it requires the specialist to use more resources, it makes it possible to construct event chains, question the relevance of the organizational variables of the reference model and specify the value to be given to these variables in order to enhance safety. As we have seen, causal analysis requires in-depth understanding of the technical processes in use at the facilities. As a result, the human factors specialists' interest in the technique and the engineers' interest in human factors should be heightened by improving or increasing the interactions between these two groups. The creation of a network of human factors correspondents within the IRSN might facilitate such exchanges.

Furthermore, the results of the causal analyses could be better shared among the human factors specialists if they were compiled more systematically. The creation of a database containing the various elements in the event chains might help to improve knowledge in the human factors assessment.

Lastly, the specialists could be encouraged to announce the results of their assessments, contribute articles to specialist publications and join academic and other networks that deal with human factors in nuclear safety.

General Conclusion

The detailed reports of three contrasting cases representative of the activity of IRSN human factors specialists led us to note the original aspect of nuclear safety assessment in this field. The analysis of this material brought to light the characteristic measures taken by the experts. These measures break down into three distinct types (rhetorical, cognitive and operational). Addressing expert assessment via this three-part classification could enhance the representations shared by researchers, professionals and maybe by citizens too, and could contribute to rethinking the organization of this activity, which our work has shown to be complex. To weigh the effectiveness of human factors assessment, it is necessary to consider the institutional context in which it evolves; its concrete repercussions are the result of ongoing technical dialogue, involving many representatives of safety institutions. In conclusion, it appears that the effectiveness of the nuclear safety external review system should be investigated in the field of human and organizational factors.

1. Enhancing Models of Expert Assessment

The three types of expert assessment action we have highlighted form one of the main results of our research. It would appear that this conception of expert assessment can be applied in other fields. It can also call into question the ways in which experts' activity is organized. As a result, certain parameters that are missing from our analysis, concerning human resources management in particular, should be given greater consideration.

1.1. Three Types of Expert Assessment Action

By underlining the collaborative approach to nuclear safety assessments and by demonstrating that, for everyone involved, it is a tool for exploring knowledge of the risks involved, the historical research referred to in the first part of this study challenged the adequacy of the canonical model. By focusing on the experts' activity, and not merely on their end reports and recommendations, doubt gave way to refutation; above all, expert assessment cannot be restricted to applying a body of established knowledge. There are actually gaps in that knowledge, since the reference organizational model is based on relationships with safety that are only plausible, not proven. However, a causal analysis is a means of reinforcing the scope of knowledge, at least "locally": the validity of the causal chains will always depend on a context which may comprise significant singular aspects, thus jeopardizing any widespread application.

To our greater surprise, we noted that human factors assessment did not, in many respects, conform to the procedural model. Human factors assessment is indeed planned and regulated by numerous mechanisms that are very much like procedures, some of which integrate the adversarial principle. It does, however, rely on extensive interaction and command of several skills which are not reflected in the procedures.

In fact, the action taken by human factors specialists falls into the following three categories:

- *rhetorical: recommend, argue and align points of view while respecting institutional constraints.*

- *cognitive: use standard analyses, create and build up links between organizational factors and safety.*

- *operational: "prompt the licensee" to act.*

These three types of action are defined by expert assessment management arrangements (scheduled meetings, review channels, approval methods, reporting), structured by interdependent effectiveness objectives, and they demand a command of knowledge specific to each type of action. Conceiving expert assessment as a series of interactions that fall within these three groups, while explaining the experts' behavior, helps to interpret the situations we encountered.

The reader may be surprised to see that these categories do not include the point of view of citizens. Several recent works on expert assessment have indeed shown that such integration is currently a key issue[275]. This exclusion of the citizen can be explained by the empirical approach we adopted; our analyses are based on data concerning the activity of experts who never considered taking citizens into account in their work, thus removing this possibility from the

---

[275] Callon, M., P. Lascoumes, et al. (2002). *Agir dans un monde incertain. Essai sur la démocratie technique*. Paris, Le Seuil, Miserey, Y. and P. Pellegrini (2006). *Le Groupe radioécologie Nord-Contentin. L'expertise pluraliste en pratique. L'impact des rejets radioactifs dans le Nord-Contentin sur les risques de leucémie*. Paris, La documentation française.

"hybrid forum" model[276]. The subject is nonetheless today a major concern for IRSN and the ASN, and the publication of expert assessment reports is currently being considered. Without addressing it directly, our work provides a glimpse of the difficulties that citizen participation in expert assessment of nuclear safety would raise: defining its exact role and the ways in which it could be integrated into the complex processes are in no way easy. By revealing to citizens the work accomplished by experts and by proposing a model that could help them understand that activity, our work may help overcome those difficulties.

1.2. Testing and Organizing the Categories

Above and beyond the specific case of human factors in nuclear safety, recommending, aligning, learning, acting and changing seem to form the bases of the work of public safety assessment agencies, the number of which is constantly on the rise[277]. Consequently, the conception of expert assessment that we propose could advantageously change the sometimes overly partial perspectives taken on a complex activity, whose subtleties we have grasped. Testing the three categories through other cases, such as another specialized area of nuclear safety chosen among those referred to as "technical", or expert assessment of other kinds of risk, would thus be an extension of our work.

Furthermore, by stressing the need for forms of organization suited to expert assessment far removed from the canonical model, we above all placed emphasis on systems of leveraging knowledge and structuring areas of competence. Researchers in management may, however, be disappointed by the omission of certain parameters, particularly the institute's career management and incentive methods. An examination of such systems and their impact on the effectiveness of expert assessment activity might help explain certain attitudes adopted by experts and thus fuel thinking on levers for action with a view to rebalancing the dimensions of effectiveness.

---

[276] Callon, M. and A. Rip (1991). *Forums hybrides et négociations des normes socio-techniques dans le domaine de l'environnement. La fin des experts et l'irrésistible ascencion de l'expertise*. Environnement, science et politique. Les experts sont formels, Paris, Germes.

[277] Chateauraynaud, F. and D. Torny (2000). *Les sombres précurseurs. Une sociologie pragmatique de l'alerte et du risque*. Paris, Editions de l'Ecole des hautes études en sciences sociales, Gilbert, C. (2000). *La mise en place d'Agences dans le domaine des risques collectifs*. Présentation de la dix-septième séance du séminaire du programme "Risques collectifs et situations de crise", Paris, Noiville, C. (2003). *Du bon gouvernement des risques*. Paris, Presses universitaires de France, Joly, P.-B. (2005). La sociologie de l'expertise scientifique: les recherches françaises au milieu du gué. *Risques, crises et incertitudes: pour une analyse critique*. O. Borraz, C. Gilbert and P.-B. Joly. Grenoble, Publications de la MSH-Alpes: 117–174 (257), Borraz, O. (2007). Les politiques du risque. *Science Politique*. Paris, Fondation nationale des sciences politiques: 288.

2. Comprehending Nuclear Safety External Review Systems

By describing the interaction of human factors specialists during the assessment process and by analyzing the results of this interaction, our work goes a step further in analyzing the theory of technical dialogue, which Cyrille Foasso regards as a specificity of French nuclear safety institutions. Investigating the effectiveness of human factors assessment naturally leads to questioning the effectiveness of the external review system, in which assessment plays a part. This evaluation must be approached carefully, however, since it would require a detailed analysis of the impact of expert assessment and probably a comparison of external review systems on an international scale.

2.1. Furthering the Theory of Technical Dialogue

While action to increase independence in nuclear safety assessment and risk control has helped to add clarity to an organization regarded as opaque, our reports have however brought to light the concrete forms and the many variations of, technical dialogue, inherited from "French cooking", which still characterizes the French risk control system.

Our study of the effectiveness of human factors assessment brought to light the benefits of such functioning. Implementing expert assessment with no limits in time avoids restricting control to its classic forms, described in particular by William Ouchi (1979), which are insufficient to grasp the human and organizational factors of nuclear safety. A second advantage of technical dialogue is the possibility of harnessing varied forms of regulation; while it is primarily based on a compliance form of regulation, we did identify some examples of regulation by deterrence. Lastly, by creating a model of knowledge production close to intervention, this technical dialogue would appear appropriate to the expert assessment of certain fields of management (training and skills, systems of accountability and delegation for example), and produces learning effects, both for the regulated parties and the regulator.

We nonetheless revealed certain limits to this technical dialogue. Firstly, when objectives consist of bringing viewpoints to a consensus and achieving compliance with institutional requirements, conditions are not favorable for furthering knowledge, which is nonetheless necessary, as we have pointed out. Secondly, restrictions on the freedom of experts, particularly in defining issues and methods of investigation, can be likened to forms of capture. The importance of these phenomena must, however, be kept in perspective: our observations do not call into question the independence of experts in expressing their judgements, as demonstrated by their focus on "safety" and by the heated debates in which they engage. The rather simplistic assumptions underlying the notion of capture (wherein the licensee manipulates experts in defense of economic interests) is undermined by our reports, which reveal more subtle phenomena; the case of skills management clearly shows that a licensee may want to improve conditions in the plant (in order to improve safety) and may use the results of experts' work for just that purpose. Lastly, there is nothing to confirm that capture phenomena are exclusive to technical dialogue, and that they would not be found in other risk control systems. We should remember that the notion of capture was conceived by Anglo-Saxons, renowned for preferring forms of regulation based on deterrence[278].

2.2. Comparing Risk Control Systems and Assessing their Effectiveness

While we were able to identify the strengths and weaknesses of an external review system underpinned by ongoing technical dialogue between licensees and regulators, an assessment of the effectiveness of such a system would require additional data. It was not possible, with our material, to list the concrete effects of expert assessment on facilities. A careful examination thereof would be an interesting extension of our work.

Moreover, a comparative analysis of nuclear risk control systems, on an international scale, would no doubt lead to including certain variables that we have not mentioned. We could give greater attention to economic parameters, as did the deputy Jean-Yves Le Déaut, by mentioning the resources of expert assessment and risk control structures (1998). This comparative analysis could focus on organizational factors, and it would indeed be interesting to compare the objects used in our analyses (the outcomes of interactions between regulator and regulated, organization of expert assessment and risk control, human and organizational variables of the reference model, standard analyses implemented, knowledge production models, etc.). Such work would no doubt be instructive for human factors expert assessment.

We must, however, emphasize the difficulties involved in assessing the effectiveness of external risk control systems. They are mainly due to the specificities of safety, as these words by a safety authority engineer illustrate: "After all, safety comes in many forms and is difficult to measure. We ride on that difficulty: we tell EDF we want more of it, without always knowing what it actually is."

---

[278] Rochlin, G. I. and A. von_Meier (1994). "Nuclear power operations: a cross-cultural perspective." *Annual review of energy and the environment*(19): 153–187, Hood, C., H. Rothstein, et al. (2004). *The government of risk. Understanding risk regulation regimes.* Oxford, Oxford University Press.

The description of the two incidents examined closely in the study further illustrates the complex combination of singular events that can threaten the safety of nuclear facilities. In one sense, the difficulty in anticipating and controlling such situations was already present in Aristotelian thinking, through the phronesis concept (prudence or practical wisdom): "The ground for prudence is ... the contingent which, when it affects us, is called *chance*"[279]. To Aristotle, prudence is neither science nor art; it is a virtue. "Virtue consists in acting according to the golden mean, and the criterion of the golden mean is the right rule. But what is the right rule? Aristotle does not give us any way of recognizing it, apart from the appeal to the judgement of the man of practical wisdom."[280] So, isn't the man of practical wisdom, the "man with a fair assessment of singular situations"[281], therefore the veritable human factor of nuclear safety?

---

[279] Aubenque, P. (1963 (2004)). *La prudence chez Aristote*. Paris, PUF, p. 30.

[280] Ibid., p. 40.

[281] Pellegrin, P. (2007). *Dictionnaire Aristote*. Paris, Ellipses, p. 165.

Postface

It is a challenge to write a postscript to a book, possibly even greater than writing a preface. The reason is that when the reader gets to the postscript, he or she will (presumably) have read the book and therefore already formed an own opinion of its strengths and weaknesses. Or, to paraphrase the title, the reader will have assessed the book. The challenge is possibly even greater in the case of this book, which is unique in both scope and contents, as well as in a refreshing non-Anglo-Saxon style of writing and arguing.

On reflection, there are two issues which I would like to comment on in this postscript. The first is the issue of the human factor – or human factors – and the other is the issue the role that a safety assessment plays.

With regard to the human factor, it is an interesting fact that human factors – or human factors engineering, as it was called in the beginning[282] – goes back to the mid-1940s. (In France, the journal *Le travail humain* started in 1937; in the UK, the *Ergonomics Research Society* was founded in 1946; and in the USA, the *Human Factors and Ergonomics Society* was founded in 1957.) Human factors was therefore not unknown when nuclear power plants came into use. But despite the fact that the experience of the U.S. Army during WW-II clearly had shown that so-called 'pilot errors' could be greatly reduced by paying attention to the design of displays and controls, human factors was not seen as being crucial for safety. The safety of nuclear power plants was assessed, but the assessment was based on PSA (Probability Safety Assessment).

This changed dramatically after the accident at TMI in 1979. As described by the book, the Kemeny report in particular pointed to the lack of operator training, the inadequacy of the emergency response procedures, and the design of the control room (displays and instrumentation). Since TMI, the importance of human factors has been undisputed. No one would today doubt that people

---

[282] Today the preferred terms seems to be Human Factors and Ergonomics (HF/E).

play a decisive role in safety, in nuclear power plants and in other high-risk industries. This is so for humans considered as individuals – the classical focus of human factors – and for humans as the anonymous collective that we call organisations. The latter rose to prominence after the disaster at Chernobyl in 1986 – and by the explosion of the space shuttle Challenger earlier the same year – and has, by analogy, become known as organisational factors.

But while there is general agreement – from licensees to regulators and lawmakers – that human and organisational factors somehow must be brought under control or regulated, the role of the human factor is neither well-defined nor is there much agreement on how this can best be achieved. As the book makes clear, the scientific knowledge is fragmented and insufficient as a reliable frame of reference, although not for a lack of trying. The orthodox approach is to gather as much data and evidence as possible, in the hope that this eventually will create a singularity of facts where quantity transforms itself to quality. While that would put assessment on a firm basis, it might not be the best solution even if it was possible. From a contemporary perspective on safety – as represented by HRO and resilience engineering – it is questionable whether linking safety problems (and solutions) too strongly with specific 'factors' is a viable approach. While it conforms with a technical or engineering ethos, it clashes with a system perspective that emphasises the need to consider systems as wholes rather than as composed of parts. If it turns out that there is no single factor or set of factors that can serve as a proxy for safety assessments, in other words if the problem is unmanageable with a traditional approach, then the focus on human factors may not be a *cul-de-sac*.

The second issue is the role of safety assessment. While the book is a rich source of information about how such assessments are done – at least within one culture – the point I want to consider here is not **how** safety assessments are done, but more **why**. The reason for this is the intriguing term in the book's subtitle, namely the 'assessment factory'.

To think of safety assessment as something that is done, or can be done, in a factory, means that it is considered as a rigorous process. Factories produce products, and the products should be as standardised as possible, to ensure both that the production process is effective and that the product is of acceptable quality.

There is, indeed, good reason to want safety assessment to be like this. The purpose of a safety assessment is obviously to confirm beyond any reasonable doubt that something is safe.[283] The assessment – the confirmation – is done by one group or entity for the benefit of another. In the case of nuclear power plants, as well as in the case of many other industrial and societal installations, the assessment is carried out by a regulator, who in this way represents the public. (To see how a similar role was played by other types of assessment or guarantee, consider two cases that caught the public eye in the beginning of 2013. One was the detection (in Europe) of horse meat in what was supposed to be – and sold as – pure beef. The other was the worldwide problems with overheating lithium batteries in the Boeing 787 Dreamliner. In both cases the safety assessments, the guarantees by the regulators to the public, were clearly not good enough.)

But in addition to the regulators acting on behalf of the public, there are other stakeholders. In the case of nuclear power plant, one stakeholder is the licensee or the power company. They need the assessment to get an operating license, but it would clearly be beneficial if the process was as simple as possible. In other words, a standardised production. Another stakeholder is the government who must rely on the regulator – or rather, on the experts – to perform the assessment competently in a way that is subject to audit.

This creates a situation for the regulators, and through them the experts, where they must achieve their objectives while at the same time satisfy multiple constraints. This is never easy to do, and the situation is further complicated by the uncertain status human factors, as argued above. The value of this book is that it provides an abundance of high quality material to enable a thorough discussion of both issues. The discussion itself will however, have to wait for another opportunity.

Erik Hollnagel
Professor, Institute of Regional Health Research, University of Southern Denmark
Chief Consultant, Centre for Quality, Region of Southern Denmark
Professor Emeritus, University of Linköping, Sweden

---

[283] This usually refers to Safety-I rather than Safety-II, i.e., safety as the prevention of accidents rather than safety as the ability to succeed. Hollnagel, E. (2013). A tale of two safeties. *International Journal of Nuclear Simulation, 4*(1).

References

Aggeri, F. (1999). "Environmental policies and innovation. A knowledge-based perspective on cooperative approaches." *Research policy*(28): 699–717.

AIEA (1996). Defence in depth in nuclear safety. INSAG-10. Vienne, IAEA: 33.

AIEA (2006). Fundamental Safety Principles safety fundamentals. Vienne, IAEA: 19.

Akrich, M., M. Callon, et al. (2006). *Sociologie de la traduction textes fondateurs*. Paris, Les presses de l'Ecole des mines.

Argyris, C. and D. A. Schön (1992). *Theory in practice: increasing professional effectiveness*. San Francisco, Jossey-Bass.

Aron, R. (1969). *La philosophie critique de l'histoire*. Paris, Librairie philosophique J. Vrin.

Aubenque, P. (1963 (2004)). *La prudence chez Aristote*. Paris, PUF.

Ayres, I. and J. Braithwaite (1992). *Responsive regulation. Transcending the deregulation debate*. New York, Oxford University Press.

Balogh, B. (1991). *Chain reaction expert debate and public participation in American commercial nuclear power, 1945–1975*. New York, Cambridge University Press.

Bancel-Charensol, L. and M. Jougleux (1997). "Un modèle d'analyse des systèmes de production dans les services." *Revue française de gestion*(113): 71–81.

Barbier, M. and C. Granjou (2003). Experts are learning. *EGOS*. Copenhaguen.

Bardach, E. and R. A. Kagan (1982). *Going by the book. The problem of regulatory unreasonableness*. Philadelphia, Temple University Press.

Barthe, Y. and C. Gilbert (2005). Impuretés et compromis de l'expertise, une difficile reconnaissance. *Le recours aux experts. Raisons et usages politiques*. L. Dumoulin, S. L. Branche, C. Robert and P. Warin. Grenoble, Presses universitaires de Grenoble: 43–62 (479).

Berry, M. (1983). Une technologie invisible? L'impact des instruments de gestion sur lévolution des sysèmes humains. Paris, Centre de recherche en gestion, Ecole Polytechnique: 98.

Berry, M., J.-C. Moisdon, et al. (1978). "Qu'est-ce que la recherche en gestion?" *Informatique et gestion* **108–109**.

Blackler, F. (1995). "Knowledge, knowledge work and organizations: an overview and interpretation." *Organization studies* **16**(6): 1021–1046.

Blockley, D. I., Ed. (1992). *Engineering safety*. Berkshire, McGraw Hill.

Blockley, D. I. (1996). Hazard engineering. *Accident and Design*. C. Hood and D. K. Jones. Abingdon, University College London Press: 31–39.

Borraz, O. (2007). Les politiques du risque. *Science Politique*. Paris, Fondation nationale des sciences politiques: 288.

Boudon, R. (1990). *L'art de se persuader des iées douteuses, fragiles ou fausses*. Paris, Librairie Artème Fayard.

Bourgeois, J. (1992). La ûreé nucéaire. *L'aventure de l'atome*. P. M. d. l. Gorce, Flammarion.

Bourrier, M. (1999). *Le nucéaireà lépreuve de l'organisation*. Paris, Presses universitaires de France.

Bourrier, M. (2007). Risques et organisations. *Face au risque*. C. Burton-Jeangros, C. Grosse and V. November. Geève, Georg: 159–182.

Boutin, P. (2001). L'expertise des facteurs de performance humaine dans les installations nucéaires, CEA/IPSN/DES/SEFH: 14.

Callon, M., P. Lascoumes, et al. (2002). *Agir dans un monde incertain. Essai sur la émocratie technique*. Paris, Le Seuil.

Callon, M. and A. Rip (1991). *Forums hybrides et égociations des normes socio-techniques dans le domaine de l'environnement. La fin des experts et l'irésistible ascencion de l'expertise*. Environnement, science et politique. Les experts sont formels, Paris, Germes.

Charron, S. and M. Tosello (1994). *Ergonomie etévaluation de ûreé dans le secteur du nucéaire*. XXIXe congès de la socéé d'ergonomie de langue fraçaise, Eyrolles.

Chateauraynaud, F. and D. Torny (2000). *Les sombres pécurseurs. Une sociologie pragmatique de l'alerte et du risque*. Paris, Editions de l'Ecole des hautesétudes en sciences sociales.

Chatzis, K., F. de_Coninck, et al. (1995). "L'accord A. Cap 2000 la "logique cométence" à lépreuve des faits." *Travail et emploi*(64): 35–47.

Chevreau, F.-R. (2008). Mîtrise des risques industriels et culture de écurié: le cas de la chimie pharmaceutique. *Sciences et énie des activiésà risques*. Paris, Mines ParisTech. **Tèse de doctorat**: 285.

Cogé, F. (1984). "Evolution de la ûreé nucéaire." *Revue éérale nucéaire*(1): 18–32.

Collingridge, D. (1996). Resilience, flexibility, and diversity in managing the risks of technologies. *Accident and design: contemporary debates in risk management*. C. Hood and D. K. C. Jones. London, UCL Press limited.

Colmellere, C. (2008). Quand les concepteurs anticipent l'organisation pour mîtriser les risques: deux projets de modifications d'installations sur deux sites clasés SEVESO 2. *Sociologie*. Paris, Universié de technologie de Compègne. **Tèse de doctorat**: 409.

Cour_des_comptes (2005). Le émanèlement des installations nucéaires et la gestion des échets radioactifs. Rapport au Pésident de la épublique suivi des éponses des administrations et des organismes inéresés. Paris, Cour des comptes: 279.

Devillers, C. (1979). L'accident de Three Mile Island. *11e congès national de l'association pour les techniques et les sciences de radioprotection*. Nantes.

Dodier, N. (1993). *L'expertise édicale. Essai de sociologie sur l'exercice du jugement*. Paris, étailé.

Dodier, N. (1994). "Causes et mises en cause. Innovation sociotechnique et jugement moral face aux accidents du travail." *Revue fraçaise de sociologie* **35**(2): 251–281.

Dodier, N. (1995). *Les hommes et les machines. La conscience collective dans les socéés techniciées*. Paris, Editions étailé.

Dupraz, B. (1986). "La prise en compte de l'exérience pour maintenir et aéliorer la ûreé des centrales nucéaires." *Annales des Mines*: 41–46.

Education_permanente (2003). "ù en est l'inénierie de la formation?: dossier." (157).

Education_permanente (2007). "Intervention et savoirs: la penée au travail." *Education permanente*(170).

Engel, F., D. Fixari, et al. (1986). Le éminaire "Pratiques d'intervention des chercheurs". *Chercheurs dans l'entreprise ou la recherche en action*. C. Alezra, F. Engel, D. Fixari and J.-C. Moisdon. Paris, Les cahiers du programme mobilisateur "Technologie, Emploi, Travail". Minisère de la recherche et de la technologie. **2**: 11–35.

Feynman, R. P. (1988). "An outsider's inside view of the Challenger inquiry." *Physics today*: 26–37.

Fixari, D., J.-C. Moisdon, et al. (1996). Former pour transformer: la longue marche des actions de requalification. Paris, Centre de gestion scientifique de l'Ecole des mines de Paris: 74.

Foasso, C. (2003). Histoire de la ûreé de lénergie nucéaire civile en France (1945–2000). Technique d'inénieur, processus d'expertise, question de socéé. *Histoire moderne et contemporaine*. Lyon, Universié Lumère – Lyon II. **Tèse de doctorat**: 698.

Fourest, B., Y. Boaretto, et al. (1980). Impact de l'accident de Three Mile Island sur le programme nucéaire fraçais et sur l'analyse de ûreé. *Conérence A.N.S./E.N.S. sur la ûreé des éacteurs thermiques*. Knoxville (U.S.A.).

Fridenson, P. (1994). "Jalons pour une histoire du centre de gestion scientifique de l'Ecole des mines de Paris. Entretiens avec Jean-Claude Moisdon et Claude Riveline." *Entreprises et Histoire* **7**: 19–35.

Gand, S. (2008). L'organisation des dynamiques de services professionnels. Logique de rationalisation, cadre de gestion et formes de colégialié. *Sciences de gestion*. Paris, Mines ParisTech. **Tèse de doctorat**.

Garbolino, E. (2008). *La éfense en profondeur: contribution de la ûreé nucéaireà la écurié industrielle*. Paris, Tec & Doc Lavoisier.

Garrick, B. J. (1992). Risk management in the nuclear power industry. *Engineering safety*. D. I. Blockley. London, McGraw-Hill: 313–346.

élard, P. (2006). Rapport sur les autoriés administratives inépendantes. Paris, Office parlementaire dévaluation de la égislation: 136.

Gilbert, C. (2000). *La mise en place d'Agences dans le domaine des risques collectifs*. Pésentation de la dix-septème éance du éminaire du programme "Risques collectifs et situations de crise", Paris.

Gilbert, C. (2002). "Risques nucéaires, crise et expertise: quel ôle pour l'administrateur?" *Revue fraçaise d'administration publique* **3**(103): 461–470.

Gilbert, C. (2005). Erreurs, éfaillances, vulérabiliés: vers de nouvelles conceptions de la écurié? *Risques, crises et incertitudes: pour une analyse critique*. O. Borraz, C. Gilbert and P.-B. Joly. Grenoble, Publications de la MSH-Alpes: 69–115 (257).

Girin, J. (1987). L'objectivation des donées subjectives. Eéments pour une téorie du dispositif dans la recherche interactive. *Qualié et fiabilié des informationsà usage scientifique en gestion*. Paris, FNEGE.

Girin, J. (1989). L'opportunisme éthodique dans les recherches sur la gestion des organisations.

Godard, O. (2003). Comment organiser l'expertise scientifique sous légide du principe de pécaution?, PREG CECO Laboratoire déconoétrie: 18.

Goldschmidt, B. (1980). *Le complexe atomique. Histoire politique de lénergie nucéaire*. Paris, Fayard.

Gomolinski, M. (1982). Laboratoire détude du facteur humain. Programme pour 1983, CEA/IPSN/DAS/LEFH: 4.

Gomolinski, M. (1985). Paraètres humains dans la ûreé des installations nucéaires, CEA/IPSN/DAS/LEFH: 5.

Gomolinski, M. (1986a). La prise en compte du facteur humain dans la conception et le fonctionnement des centralesà eau éère, CEA/IPSN/DAS/LEFH.

Gomolinski, M. (1986b). Paraètres humains dans la ûreé des installations nucéaires, CEA/IPSN: 10.

Granjou, C. (2004). La gestion du risque entre technique et politique. Comiés d'experts et dispositifs de trçabiliéà travers les exemples de la vache folle et des OGM. *Sociologie*. Paris, Universié Paris 5 Reé Descartes. **Tèse de doctorat**: 488.

GRETU (1980). *Uneétudeéconomique a monté ... Mythes et éaliés sur lesétudes de transports*. Paris, Editions Cujas.

Gérin, F., A. Laville, et al. (1997). *Comprendre le travail pour le transformer. La pratique de l'ergonomie*. Lyon, Editions de l'ANACT.

Hatchuel, A. (1994a). "Les savoirs de l'intervention en entreprise." *Entreprises et Histoire*(7): 59–75.

Hatchuel, A. (1994b). "Apprentissages collectifs et activiés de conception." *Revue fraçaise de gestion*(99).

Hatchuel, A. (2000). Quel horizon pour les sciences de gestion? Vers une téorie de l'action collective. *Les nouvelles fondations des sciences de gestion. Eéments dépisémologie en management*. A. David, A. Hatchuel and R. Laufer. Paris, Vuibert.

Hatchuel, A. (2001). "The two pillars of new management research." *British journal of management* **12**: 33–39.

Hatchuel, A. and B. Weil (1992). *L'expert et le sysème*. Paris, Economica.

Hecht, G. (2004). *Le rayonnement de la France. Energie nucéaire et identié nationale aprèS la seconde guerre mondiale*. Paris, La écouverte.

Hermitte, M.-A. (1997). "L'expertise scientifiqueà finalié politique. éflexions sur l'organisation et la responsabilié des experts." *Justices*(8): 79–103.

Hollnagel, E. (2006). Resilience – the challenge of the unstable. *Resilience engineering*. E. Hollnagel, D. D. Woods and N. Leveson. Hampshire, Ashgate: 9–17.

Hollnagel, E. and E. Rigaud, Eds. (2006). *Proceedings of the second resilience engineering symposium. 8–10 November 2006, Antibes*. Paris, Les Presses de l'Ecole des Mines.

Hollnagel, E., D. D. Woods, et al. (2006). *Resilience engineering*. Hampshire, Ashgate.

Hood, C. and D. K. C. Jones, Eds. (1996). *Accident and design. Contemporary debates in risk management*. Abingdon, University College London Press.

Hood, C., H. Rothstein, et al. (2004). *The government of risk. Understanding risk regulation regimes*. Oxford, Oxford University Press.

Houé, C. and J.-M. Oury (1981). "L'importance de la fiabilié humaine pour la ûreé des installations nucéaires. L'exérience fraçaise et les enseignements de l'accident de Three Mile Island." *Revue éérale nucéaire*(5): 419–423.

Joly, P.-B. (2005). La sociologie de l'expertise scientifique: les recherches fraçaises au milieu du gé. *Risques, crises et incertitudes: pour une analyse critique*. O. Borraz, C. Gilbert and P.-B. Joly. Grenoble, Publications de la MSH-Alpes: 117–174 (257).

Jouré, B. (1999). Les organisations complexesà risques: érer la ûreé par les ressources. Etude de situations de conduite de centrales nucéaires. *Sciences de l'homme et de la socéé. Sécialié Gestion*. Paris, Ecole Polytechnique. **Tèse de doctorat**: 434.

Keller, W. and M. Modarres (2005). "A historical overview of probabilistic risk assessment development and its use in the nuclear power industry: a tribute to the late Professor Norman Carl Rasmussen." *Reliability engineering and System safety*(89): 271–285.

Kemeny, J. (1979). Report of the President's Commission on the accident of Three Mile Island, www.pddoc.com/tmi2/kemeny

La_Porte, T. R. and P. M. Consolini (1991). "Working in practice but not in theory: theoritical challenges of "high-reliability organizations"." *Journal of Public administrations*(1).

La_Porte, T. R. and C. W. Thomas (1995). "Regulatory compliance and the ethos of quality enhancement: surprises in nuclear power plant operations." *Journal of public administration research and theory* **5**(1): 109–137.

Lagadec, P. (1981). *Le risque technologique majeur. Politique, risque et processus de éveloppement*. Paris, Pergamon Press.

Lagadec, P. (1988). *Etats d'urgence. éfaillances technologiques et éstabilisation sociale*. Paris, Seuil.

Le_éaut, J.-Y. (1998). Rapport sur le sysème fraçais de radioprotection, de contôle et de écurié nucéaire: la longue marche vers l'inépendance et la transparence.

Leclerc, O. (2005). *Le juge et l'expert. Contributionà létude des rapports entre le droit et la science*. Paris, LGDG.

Lelèvre, J. (1974). "L'analyse de ûreé et lesétudes correspondantes." *Annales des Mines*(Janvier): 55–61.

Libmann, J. (1996). *Eéments de ûreé nucéaire*. Les Ulis, Leséditions de physique.

Llory, M. (1999). *L'accident de la centrale nucéaire de Three Mile Island*. Paris, L'Harmattan.

Llory, M. (2000). *Accidents industriels: le cût du silence. Oérateurs priés de parole et cadres introuvables*. Paris, L'Harmattan.

Mazuzan, G. T. and J. S. Walker (1985). *Controlling the atom*. Berkeley, University of California press.

Millet, A.-S. (1991). L'invention d'un sysème juridique: Nucéaire et Droit. *Droit*. Nice, Universié de Nice-Sophia Antipolis. **Tèse de doctorat**: 625.

Mintzberg, H. (1978 (2005)). *Structure et dynamique des organisations*. Paris, Editions d'Organisation.

Mintzberg, H. (1980). "Structure in 5's: a synthesis of the research on organization design." *Management science* **26**(3): 322–341.

Miserey, Y. and P. Pellegrini (2006). *Le Groupe radiécologie Nord-Contentin. L'expertise pluraliste en pratique. L'impact des rejets radioactifs dans le Nord-Contentin sur les risques de leuémie.* Paris, La documentation fraçaise.

Moatti, J.-P. (1989). *Economie de la écurié: de lévaluationà la pévention des risques technologiques.* Paris, La documentation fraçaise.

Moisdon, J.-C. (1984). "Recherche en gestion et intervention." *Revue fraçaise de gestion*: 61–73.

Moisdon, J.-C., Ed. (1997). *Du mode d'existence des outils de gestion.* Paris, Seli-Arlsan.

Noiville, C. (2003). *Du bon gouvernement des risques.* Paris, Presses universitaires de France.

Nonaka, I. and H. Takeuchi (1995). *The knowledge-creating company: how japanese companies create the dynamics of innovation.* Oxford, Oxford University Press.

Ouchi, W. G. (1979). "A conceptual framework for the design of organizational control mechanisms." *Management science* **25**(9): 833–848.

Oudiz, A. and G. Doniol-Shaw (2005). Histoire de l'ergonomie dans les entreprises: l'IPSN. *Bulletin de la socéé d'ergonomie de langue fraçaise.* Paris.

Oudiz, A., E. Guyard, et al. (1990). Gestion de la fiabilié humaine dans l'industrie nucéaire, quelquesééments. *Les facteurs humains de la fiabilié dans les sysèmes complexes.* J. Leplat and G. d. Terssac. Marseille, Octaès: 273–292.

Paé-Cornell, E. (1990). "Organizational aspects of engineering system safety: the case of offshore platforms." *Science* **250**: 1210–1217.

Pellegrin, P. (2007). *Dictionnaire Aristote.* Paris, Ellipses.

Peltzman, S. (1980). "The growth of government." *Journal of law and economics* **23**: 209–287.

Perelman, C. (1977 (2002)). *L'empire rétorique. Rétorique et argumentation.* Paris, Librairie philosophie J. Vrin.

Perelman, C. and L. Olbrechts-Tyteca (1958 (2000)). *Traié de l'argumentation.* Bruxelles, Editions de l'Universié de Bruxelles.

Perin, C. (1998). "Operating as experimenting: synthesizing engineering and scientific values in nuclear power production." *Science, technology & human values* **23**(1): 98–128.

Perrow, C. (1983). "The organizational context of human factors engineering." *Administrative science quaterly* **28**: 521–541.

Perrow, C. (1984 (1999)). *Normal accidents. Living with high-risk technologies.* Princeton, Princeton University press.

Perrow, C. (1994). "The limits of safety: the enhancement of a theory of accidents." *Journal of contingencies and crisis management* **2**(4): 212–220.

Plot, E. (2007). *Quelle organisation pour la mîtrise des risques industriels majeurs écanismes cognitifs et comportements humains.* Paris, L'Harmattan.

Raush, J.-M. and R. Pouille (1987). Conéquences de l'accident de la centrale nucéaire de Tchernobyl et ûreé et écurié des installations nucéaires, Office parlementaire dévaluation des choix scientifiques et techniques.

Reason, J. (1987). "The Chernobyl errors." *Bulletin of the British psychological society*(40): 201–206.

Reason, J. (1990). "The age of organizational accident." *Nuclear engineering international* 18–19.

Reason, J. (1993). *L'erreur humaine*. Paris, Presses universitaires de France.

Reason, J. (1997). *Managing the risks of organizational accidents*. Aldershot, Ashgate Publishing Limited.

Reason, J. (2006). Human factors: a personal perspective. Helsinki.

Reason, J., E. Hollnagel, et al. (2006). Revisiting the "swiss cheese" model of accidents. Bétigny-sur-Orge, Eurocontrol.

Rees, J. V. (1994). *Hostages of each other: the transformation of nuclear safety since Three Mile Island*. Chicago, The University of Chicago Press.

Riveline, C. (1991). "Un point de vue d'inénieur sur la gestion des organisations." *érer et comprendre*: 50–62.

Roberts, K. (1990). "Some characteristics of one type of high reliability organization." *Organizations science* **1**(2): 160–176.

Rochlin, G. I. (1999). The social construction of safety. *Nuclear safety. A human factors perspective*. J. Misumi, B. Wilpert and R. Miller. London, Taylor & Francis: 5–23.

Rochlin, G. I. (2001). les organisations à haute fiabilité: bilan et perspectives de recherche. *Organiser la fiabilié*. M. Bourrier. Paris, L'Harmattan: 39–70.

Rochlin, G. I. and A. von_Meier (1994). "Nuclear power operations: a cross-cultural perspective." *Annual review of energy and the environment*(19): 153–187.

Rolina, G. (2005). Savoirs et discours de l'expert. Le cas du sécialiste des facteurs humains de la ûreé nucéaire. *14e rencontres "Histoire & Gestion"*. Toulouse.

Rolina, G. and L. Roux (2006). La recherche-intervention en gestion. Une ééalogie de l'intervention au CGS. *Intervenir dans le monde du travail: la responsabilié sociale d'un centre de recherche en sciences humaines*. Lège.

Roqueplo, P. (1995b). "Scientific expertise among political powers, administrations and public opinion." *Science and Public Policy* **22**(3): 175–182.

Roqueplo, P. (1997). *Entre savoir et écision, l'expertise scientifique*. Paris, INRA Editions.

Sagan, S. D. (1993). *The limits of safety. Organizations, accidents and nuclear weapons*. Princeton, Princeton studies in international history and politics.

Schmetz, R. (2000). *L'argumentation selon Perelman: pour une raison au coeur de la rétorique*. Namur, Presses universitaires de Namur.

Shrivastava, P. (1987). *Bhopal: anatomy of a crisis*. Cambridge, Ballinger.

Simonnot, P. (1978). *Les nucéocrates*. Grenoble, Presses universitaires de Grenoble.

Sparrow, M. K. (2000). *The regulatory craft*. Washington, The brookings institution.

Stigler, G. J. (1971). "The theory of economic regulation." *Bell journal of economics and management sciences* **21**: 3–21.

Tasset, D. (1998). "Palier N4:évaluation de la ûreé des aspects facteurs humains de la salle de commande informatiée." *Revue éérale nucéaire*(1): 20–26.

Theureau, J. and F. Jeffroy (1994). *Ergonomie des situations informatiées. La conception centée sur le cours d'action de l'utilisateur*. Toulouse, Octaès.

Tépos, J.-Y. (1996). *La sociologie de l'expertise*. Paris, Presses universitaires de France (coll. Que sais-je?).

Turner, B. A. (1978). *Man-made disasters*. London, Wykeham.

Vallet, B. (1984). La ûreé des éacteurs nucéaires en France: un cas de gestion des risques. Rapport au service central de ûreé des installations nucéaires, Ecole des mines de Paris – Centre de sociologie de l'innovation: 123.

Vallet, B. (1985). *La constitution d'une expertise de ûreé nucéaire en France*. Situations d'expertise et socialisation des savoirs, Saint-Etienne.

van_Eemeren, F. and R. Grootendorst (1996). *La nouvelle dialectique*. Paris, Editions Kié.

van_Eemeren, F. H. and P. Houtlosser (2004). Une vue synoptique de l'approche pragma-dialectique. *L'argumentation aujourd'hui. Positions téoriques en confrontation*. M. Doury and S. Moirand. Paris, Presses Sorbonne Nouvelle: 45–75.

Vaughan, D. (1990). "Autonomy, interdependence, and social control: NASA and the space shuttle Challenger." *Administrative science quaterly* **35**: 225–257.

Vaughan, D. (1996). *The Challenger launch decision: risky technology, culture, and deviance at Nasa*. Chicago, University of Chicago Press.

Veyne, P. (1971 (1996)). *Comment onécrit l'histoire?* Paris, Seuil.

Veyne, P. (1983). *Les Grecs ont-ils cruà leurs mythes?* Paris, Seuil.

Vidaillet, B., Ed. (2003). *Le sens de l'action. Karl E. Weick: sociopsychologie de l'organisation*. Paris, Vuibert.

Villemeur, A. (1988). *ûreé de fonctionnement des sysèmes industriels*. Paris, Eyrolles.

Walker, J. S. (2004). *Three Mile Island: a nuclear crisis in historical perspective*. Berkeley, The University of California Press.

Weick, K. (1976). "Educational organizations as loosely coupled systems." *Administrative science quaterly* **21**(1): 1–19.

Weick, K. (1987). "Organizational culture as a source of high reliability." *California management review* **29**(2): 112–127.

Weick, K. and K. Sutcliffe (2001). *Managing the unexpected*. San Fransisco, Jossey-Bass.

Wildavsky, A. (1988). *Searching for safety*. New Brunswick, Transaction publishers.